Corruption and Public Administration

T0199706

Corruption and Public Administration looks at public-sector organizations and what they have achieved since signing the United Nations Convention Against Corruption (UNCAC) agreement in Merida in 2003. It examines how the signee countries engaged in the setup of institutions to contain corruption in public administration, and how these governments and institutions have progressed. The book compares several developed countries and undertakes an especially detailed examination of Italy. It highlights strengths and weaknesses, and proposes organizational means of addressing issues which include diversity in organizational structures and systems, and a focus on prevention rather than repression.

The book shines a light on anti-corruption practices and aims to foster open discussion about this pressing topical issue among peers in all relevant fields of the social sciences.

Francesco Merloni is a member of the governing body of the Italian Anti-corruption Authority (ANAC). He was a professor of administrative law at the University of Perugia, Italy.

Corruption and Public Administration

The Italian Case in a Comparative Perspective

Francesco Merloni

Routledge
Taylor & Francis Group

LONDON AND NEW YORK

First published 2019
by Routledge
2 Park Square, Milton Park, Abingdon, Oxon OX14 4RN

and by Routledge
605 Third Avenue, New York, NY 10017

First issued in paperback 2020

Routledge is an imprint of the Taylor & Francis Group, an informa business

British Library Cataloguing-in-Publication Data
A catalogue record for this book is available from the British Library

Library of Congress Cataloging-in-Publication Data
A catalog record has been requested for this book

ISBN 13: 978-0-367-73335-3 (pbk)
ISBN 13: 978-1-138-36672-5 (hbk)

Typeset in Bembo
by Wearset Ltd, Boldon, Tyne and Wear

Contents

Exhibits

About the author

Francesco Merloni is a member of the governing body of the Italian Anti-corruption Authority (ANAC). He was professor of administrative law at the University of Perugia, Italy, and has been a member of several committees on corruption prevention in the Italian Ministry for Public Administration:

- Member of the anti-corruption working group, Ministry of Public Administration;
- Member of the scientific committee of the project on measures to contain corruption in local and central public administration;
- Member of the working group on conflict of interest in public appointments, Ministry of Public Administration'
- Member of the working group on transparency, Ministry of Public Administration.

He is the author of, among others:

Merloni, F. (2006). Dirigenza pubblica e amministrazione imparziale: Il modello italiano in Europa. *Il Mulino*.
Merloni, F. (Ed.) (2008). *La trasparenza amministrativa*. Milan: Giuffré.
Merloni, F. (2016). *Istituzioni di diritto amministrativo*. Turin: Giappichelli Editore.

Acknowledgements

I gratefully acknowledge helpful comments from Lucio Bianco, Paolo D'Anselmi, Daniele David, Alessandro Ferrara, Emma Galli, Andrea Lapiccirella and two anonymous referees. I would also like to thank Dean Bargh, Athanasios Chymis, Toni Muzi Falconi, Daniela Greco, Matthew LePere, John Paluszek, Paul Parks, Maria Grazia Quieti, Vittorio Scaffa and Daniel Waterman.

Acronyms and abbreviations

Abbreviation	In full	Translation
AFA	Agence Française Anticorruption	French Anti-corruption Agency
AGCM	Autorità Garante della Concorrenza e del Mercato	Italian Competition (antitrust) Authority
Agenas		National Agency for Health Care
ANAC	Autorità Nazionale Anticorruzione	National Anti-corruption Authority (Italy)
ANAS	Azienda Nazionale Autonoma Strade	National Road Construction and Maintenance Company
ANCE	Associazione Nazionale Costruttori Edili	Building contractors' association
ANCI	Associazione Nazionale Comuni Italiani	
ARAN	Agenzia Rappresentanza Negoziale Pubbliche Amministrazioni	Negotiating Agency for Public Administrations
ATAC	Agenzia del trasporto autoferrotranviario del Comune di Roma	Rome transport company
AVCP	Autorità per la Vigilanza sui Contratti Pubblici di Lavori, Servizi e Forniture	Authority for the Supervision of Public Contracts
CIVIT	Commissione per la Valutazione, l'Integrità e la Trasparenza delle pubbliche amministrazioni	Committee on Evaluation, Integrity and Transparency of Public Administration
CPI	Corruption Perception Index	
CRUI	Conferenza dei Rettori dell'Università Italiana	Association of the Italian Universities' Rectors

CSR	Corporate Social Responsibility	
CUG	Comitato Unico di Garanzia	Committee of Guarantors
d.l	Decreto Legge	Law Decree
d.lgs	Decreto Legislativo	Legislative Decree
DFP	Dipartimento della Funzione Pubblica	Department of Public Service
d.p.r	Decreto del Presidente della Repubblica	Decree of the President of the Republic
EPRS	European Parliamentary Research Service	
GDP	gross domestic product	
GRECO	Group of States against Corruption (European Council)	
GRI	Global Reporting Initiative	
ICAC	Independent Commission Against Corruption	
INAIL	Istituto Nazionale Assicurazione Infortuni sul Lavoro	National Insurance for Work Safety
INPS	Istituto Nazionale Previdenza Sociale	National Institute for Social Security
IRG	Implementation Review Group	
ISTAT		Italian National Institute of Statistics
l.	Legge	Law, Public Law
MEF	Ministero Economia e Finanze	Ministry of Economy and Finance
MISE	Ministero Industria e Sviluppo Economico	Ministry of Industry and Economic Development
MIUR	Ministero dell'Istruzione, dell'Università e della Ricerca Scientifica	Ministry for Education, University and Scientific Research
OE	economic operator	Enterprise, business firm
OECD	Organization for Economic Co-operation and Development	
OCLCIFF	Office Central de Lutte contre la Corruption et les Infractions Financières et Fiscales	Central Office for the Struggle against Corruption and Financial and Fiscal Infractions (France)

OIV	Organismi Indipendenti di Valutazione della performance	Independent Bodies for Performance Evaluation
ONG	Organizzazioni Non Governative	Non-governmental organization (NGO)
PNA	Piano Nazionale Anticorruzione	National Anti-corruption Plan
PP	Piano della Performance	Performance Plan
PTF	Piano Triennale di Formazione	Three-Year Plan for Training
PTTI	Piano Triennale per la Trasparenza e l'Integrità	Three-Year Plan for Transparency and Integrity
PTPCT	Piano Triennale per la Prevenzione della Corruzione e la Trasparenza	Three-Year Plan for Corruption Prevention and Transparency
RPC	Responsabili di Prevenzione della Corruzione	Corruption Prevention Officers
SAeT	Servizio Anticorruzione e Trasparenza	Anti-Corruption and Transparency Service
SCPC	Service Central de Prévention de la Corruption	Central Service for Prevention of Corruption (France)
SDGs	UN Sustainable Development Goals	
SDSN	UN Sustainable Development Solutions Network	
SFO	Serious Fraud Office, UK	
SNA	Scuola Nazionale di Amministrazione	National School of Administration
SOE	state-owned enterprise	
SSN	Servizio Sanitario Nazionale	National Health Service
UNCAC	United Nations Convention Against Corruption	
UNGC	United Nations Global Compact	
UPD	Ufficio Procedimenti Disciplinari	Internal Discipline Bureau
UPI	Unione Province Italiane	Union of Italian Provinces
USOGE	United States Office of Government Ethics	
WGB	Working Group on Bribery (OECD)	

Introduction

There is a widely shared perception that corruption in modern democracy is a pervasive and systemic phenomenon which affects society as a whole. Indeed, behaviour associated with corruption undermines confidence in public institutions, distorts fair competition between companies, causes an enormous increase in average costs (and in delays) of infrastructure, leads to poor quality in public works (Della Porta and Vannucci, 2005) and constitutes an economic and social burden for a country (Johnston, 2012; Rose-Ackerman, 1996; Gray and Kaufman, 1998; Thompson, 1992). In response to this problem, several countries have launched policies to tackle corruption based not on a purely repressive approach, but from a perspective of prevention and containment of the risk of corruption.

This book is the account of some countries' efforts to contain corruption in public administration following the United Nations Convention Against Corruption (UNCAC), the Merida Convention of 2003. The UNCAC was reached under the auspices of the UN Office on Drugs and Crime, in Vienna, whereby signatory countries agreed to set up organizations and institutions to address corruption in public administration. This book takes this agreement as its starting point. It uses a comparative approach which examines several countries and identifies different possible organizational ways of approaching the issue; for instance, different organizational structures and systems are examined. The book also looks at differing social goals of the anti-corruption effort (e.g. prevention vs. repression). The emphasis is on prevention rather than repression. The implementation approach looks at one country (Italy) in more depth and shows in detail how that country is working, on a day-to-day basis, to maintain its commitment to the Merida agreement. In fact, as we will see, its implementation can bring about unintended consequences, some beneficial.

To obtain a global perspective, this book will look at several European Organization for Economic Co-operation and Development (OECD) countries (France, Italy and the UK) as well as some non-European OECD ones (the US and Australia). This selection also stems from an interest in the more recent effects of the UNCAC of 2003, viewed over a longer time-horizon, by which measure countries were also chosen as a function of their historical or recent establishment of anti-corruption institutions.

It is important to point out at this stage that, first, current concerns about corruption are largely directed towards developing countries. This study, focused as it is on the so-called industrialized (and "tertiarized") world, might serve as a reminder that corruption is not only an issue connected to economic development. The industrialized countries are by no means immune to the effects of corruption, although our study identifies different causes and potentially different remedies—remedies that might nonetheless also be applicable in conditions of economic development.

Second, the literature appears to give evidence of only sectoral or institution-specific efforts to combat corruption, whereas we bring a broader focus to bear, looking at nationwide approaches including central—and, in several cases, regional and local—government. Our approach is also comprehensive by nature, including, indeed fostering, organizational and legislative change.

Third, we ask the question: how do countries and governments develop an anti-corruption initiative in the first place? This book takes account of the stakeholders of anti-corruption activity and examines the sources of demand for good government—something that has not attracted authors' attention thus far. Included among these stakeholders are critics of the anti-corruption effort. Stakeholders also include the international and supranational bodies that put pressure on national governments and contribute comparative theory and practice.

Finally, we can differentiate this study from existing literature in the following ways: (a) an understanding of the self-management and decentralized responsibility of individual public administration organizations as opposed to the complete responsibility of the anti-corruption organization; (b) the inclusion of state-owned enterprises (SOEs) within the ambit of anti-corruption action; and (c) a focus on prevention rather than repression (implying no desire to substitute for judicial action, however).

This book will follow an implementation approach. An implementation approach is characterized by a detailed description of the working of an organization, from statutory requirements to day-to-day operations. It is also characterized by a quest for concrete results and an impact on effectual reality, acknowledging that the practicalities of implementation can frustrate the outcomes of even the best legislation. The approach also remains aware of the economic and organizational environment in which an anti-corruption organization is called upon to act. This is a holistic view, in pursuit of the well-being of the country as a whole.

The implementation approach foregrounds a specific country, but also places that country's economy within its international context and with regard to its ranking by international organizations such as the World Bank.

The specific country in question is Italy. Corruption in Italy is a major problem that has over time come to characterize the political and administrative system (Newell and Bull, 2003). News coverage alone speaks to the critical nature of the corruption situation in Italy and, according to Transparency International (2018), Italy ranks 61st among 168 countries whereas the other countries of Western Europe are usually among the top 20. For example, Denmark is first, followed by the other Scandinavian countries, the Netherlands fifth, and the UK and Germany tenth. Eurobarometer (European Commission, 2012) paints a picture of a country with major problems in terms of legality and public ethics, albeit with very considerable regional differences (Quality of Government Institute, 2010), related to the concentration of organized crime (Vannucci, 2013).

As far as corruption and government effectiveness are concerned, Italy finds itself in the middle ground between developed and developing countries. It holds this position by virtue of its size, population, the value of its economy, employment levels, and by the extent of its public administration and the wider public sector, owing to the widespread existence of SOEs that are within the authority's purview. The depressing effect of corruption on the economy is estimated to be one of the highest in terms of "foregone" GDP.

Italy also represents a unique experiment in the distinction between politics and administration, via the rigorous definition of their discrete roles and tasks: it is for the political authorities to make public policies, but it is up to the administration to carry them out (Merloni, 2006). Consequently, the public administration must be assessed on its results in implementing policies, on the quality and quantity of goods and services provided to the population, not on its willingness to bend to the interests and partisan logic of the political class.

The implementation approach also helps us bridge the public and private domains, as we can call upon international corruption initiatives that have been happening separately on both sides for several years. In fact, anticorruption reform movements in public administration and in the private sector appear to be now converging in their practice. As mentioned, the public administration effort was sanctioned by the UN Convention Against Corruption (UNCAC) of 2003. The private-sector effort was sanctioned by the UN Global Compact (UNGC) Principle #10, added in 2004 to its initial nine principles of 2000. This book shows how and why this convergence is happening and how it can be understood from a theoretical point of view. We argue here that the most recent manifestations of the UNCAC are technically very similar to those of the UNGC.

One objective of this work is to provide an international audience with material and an opportunity for reflection. Our imagined reader is a professional in an international organization or an international bank, a consultant to business or public administration, or an academic. Alternatively, he or she might be a professional working in any country of the world who is one of the over half a billion public administration employees globally. Any discussion that this work stimulates will hopefully be of use to anti-corruption organizations themselves. It remains true that their actions and institutions appear to constitute an experiment (Cantone, 2016), the results of which are still uncertain or are possibly incapable of being measured.

Another objective of this work is to bring a country-specific case into the wider international arena to promote the notion of open government and to cultivate discussion among peers in all fields of the relevant social sciences as a key principle for development. In fact, openness of government is a goal best pursued through outgoing rather than incoming information: international collaborations notwithstanding, public administrations tend to be autarkic with their development constrained within their own country's boundaries, limited by an absence of plurality, language, history and culture. This self-contained nature of public administration is particularly noticeable when compared to industry, where comparison and competition make up the essence of its development dynamics.

This book is divided into two parts. The first follows a comparative approach on a country-by-country basis, looking specifically at public anti-corruption institutions. The second part continues the comparisons, but this time from the global perspective of UN efforts to address corruption, as seen from private enterprises' and public institutions' points of view. In seeking to gain an understanding of the overall effectiveness of an anti-corruption effort, the first part takes an organizational and legal comparative perspective, whereas the second takes a socio-economic perspective.

Our overriding aim in Part I, titled "The international landscape of anti-corruption efforts", is to retain a focus on the effective capabilities of organizations stemming from the Merida Convention. The narrative attempts to stick to the basic logic behind the actions of these public anti-corruption organizations as they interact with public and private organizations and institutions.

Chapter 1, "Varieties of approach and country cases", describes the anti-corruption efforts of five countries: France, Italy, the UK, the USA and Australia. Descriptions are given of the specific characteristics of the public anti-corruption organizations, such as their institutional setup, their constitutional independence from the branches of power, the approach they take (prevention or repression), and the scope of their operations within the country's public and private sectors. Chapter 2, "The mission and operations of the Italian experiment", includes an in-depth description of the anti-corruption effort in Italy. It describes the Italian Anti-Corruption Authority's main flow of action, which actively involves the country's entire public sector, including public administration and SOEs. It arrives at a description of the basic mechanism—and the theory of change thereby implied—via an excursus on the Authority's statutory and organizational history since 2009. In investigating the Authority's basic operating concept, further tasks and operations are identified, which are illustrated in Chapter 3, "An inside-out view of Italy's Anti-corruption Authority". This chapter details the Italian

Authority's inner workings and offers a perspective on possible future strategies. Chapter 4, "Country comparison of anti-corruption efforts", draws some conclusions from the investigations of individual institutions. Common traits are identified, as are limits to the current approach. Specific instruments for corruption prevention are highlighted.

Part II, titled "Broadening the view: adding contextual elements", widens the scope of the analysis. Chapter 5, "The social and economic context of the anti-corruption effort", presents the entire universe in which the Italian Authority operates, from a general view of the country, to public administration, including SOEs and other organizations that are subject to the Authority's supervision. More contextual information is provided on comparable countries. Details are also given about public expenditure and the structure of the public administration. Chapter 6, "Measuring the impact of the anti-corruption effort", deals with ways to measure the outcome of the Authority's actions, looking at the diverse types of corruption impact: from individual corrupt transactions, to the impact on government operations, to the depressing effect on GDP (foregone GDP) and the general tone of the country's social climate. Chapter 7, "Origin and support of the anti-corruption effort: the stakeholders", describes social groups and international organizations—both explicit and implicit—engaged in the Authority's activities, including the actual *creation* of the anti-corruption effort. Chapter 8, "Corruption case histories", offers specific case studies of where the Authority has taken action and of judicial action on corruption that occurred in the years under observation.

Chapter 9, "Joining public and private anti-corruption efforts", offers findings from the analyses in Part II and once again steps back to view the international context. The efforts of the UN within both the public and private sectors of the economy are compared and a proposal is made to promote the flourishing of the anti-corruption environment. Finally, in Chapter 10, the book presents some preliminary conclusions based on our findings and lists some directions for future work.

The book closes with an interview with the Italian Anti-corruption Authority's President. Here, further issues are raised and examined, and some ideas about potential solutions are tendered from a top management point of view.

The book also includes a glossary of technical terms specific to anti-corruption, which we hope can prove useful in facilitating international dialogue.

Part I

The international landscape of anti-corruption efforts

Chapter 1

Varieties of approach and country cases

1.1 The nature of corruption

Corruption is not automatically associated with developed countries. However, to imagine that corruption is exclusive to developing countries would be wrong, and developed countries also need to face up to it, albeit on a different scale. The situation implies that there are causes yet to be explored and solutions yet to be put on the table. This study focuses on developed countries and aims to play a part in efforts to address corruption, even if the expectations are as ambitious as those expressed by Kofi Annan: "to eliminate the scourge of corruption from the face of the Earth".[1]

Corruption is unfortunately a widespread phenomenon. International rankings, based on citizens' perceptions (Carloni, 2017), paint a bleak a picture which the empirical experience from daily life only serves to confirm. This is especially true when corruption is defined in a broader sense to include "maladministration". In developed countries, it is a phenomenon that has been underestimated for too long: even in reports by public bodies as recent as the early 2000s, the very existence of corruption was questioned and public concern was attributed to biased media reports.

Today, the underplaying of corruption is, at least to some extent, behind us, not least because of the 2003 UNCAC. Awareness has been raised that the damage it causes goes beyond the scope of any single public contract or individual action. Corruption has a far-reaching impact on society, as stated in this book's introduction. It undermines government effectiveness and the capacity of the state to fulfil its citizens' expectations of economic and social governance. Indeed, containment of corruption can be seen as an end in itself.

In examining the anti-corruption efforts of developed countries, this study assesses their current status in order to speculate about the future of such efforts, especially by those countries—such as France and Italy—that have begun their journey as a consequence of the UNCAC. With this study's overriding comparative approach in mind, this chapter begins by describing the landscape of a handful of countries' approach to corruption. In order to do this, we begin by identifying a set of countries to be studied. This is followed by the definition of a framework within which each country's effort is going to be analysed and described. The subsequent core of the chapter is a discussion of cases of corruption containment on an individual country basis.

1.2 The set of countries under observation

The first step in this chapter is to identify which countries comprise our sample. Since our focus is developed countries, and Italy in particular, we are looking for countries of comparable size and with similar institutions. As such, France and the UK were selected: both are European and both are similar to Italy, from a worldwide perspective. To position this sample in a global framework, we chose two further countries: the USA and Australia. The USA was chosen because it has a long record of institutional efforts towards corruption containment and because it also represents a benchmark from an academic point of view. Australia was chosen for its positive record of corruption containment but also because it is smaller than the USA (in population, GDP, etc.), and in this regard closer to the three European countries. All five countries are members of the Organization for Economic Co-operation and Development (OECD). Three are of Anglo-Saxon culture and two of a Latin culture.

Variations in the countries' timelines should be noted. The USA's initiative began in the 1970s, well before the Merida Convention (aka the UNCAC); the UK's and Australia's in the early 2000s, in parallel with the Convention's initial steps; Italy started in 2009 and accelerated in 2012; France started in 2016.

It is not our intention to attach any statistical significance to this selection of countries. A blended method of document analysis, comparative analysis and case study analysis has been employed in this chapter.

1.3 The dimensions of the country analysis

This section outlines a framework for analysis of each country in pursuit of comparable information. The chapter will go on to focus on one public anti-corruption organization in each of the observed countries. Each organization will be described according to seven dimensions:

1   The institutional setup;
2   The scope of the organization;
3   The definition of corruption being used;
4   The character of the anti-corruption effort (preventative or repressive);
5   The anti-corruption organization's relationship with the actual adminis-trative situation (macro vs. micro);
6   The focus of its measures (objective vs. subjective);
7   The relationship between the anti-corruption effort and the govern-ment's transparency effort.

Below we describe each dimension in more detail.

1.3.1 The institutional setup

The fundamental information about each organization is its degree of independence from the branches of power. Thereafter there is a brief discussion of its organizational arrangements, followed by an illustration of its *modus operandi*.

The independence from the branches of power of a specialized anticorruption institution is a fundamental requirement for the effective exercise of its functions (OECD, 2008–2013). The basic question to ask is whether the organization is an independent agency or whether it is positioned within the executive branch. These two possibilities are not exhaustive but they cover most cases. When an anti-corruption organization is positioned within the executive branch, it is seen to be less independent than when it is autonomous from all three branches. However, the details of the organizational arrangements must be taken into account. The advantages of independence are clear, but it would be interesting to identify any advantages of non-independence.

Organizational arrangements refer not only to the anti-corruption organization's structure. They also concern the systems that it uses in pursuit of its own mission, and the relationship between the anti-corruption organization and the organizations it supervises. One specific item to look for is the nature of the link between the anti-corruption organization and "the field", i.e. public administration and private business organizations.

The anti-corruption organizations will also be described according to their *modi operandi*. How do they imagine they are going to contain corruption? What is the underlying anti-corruption strategy in their mission?

1.3.2 The scope of the organization

The breadth of an anti-corruption organization's mandate is of interest. The key question is: which organizations are subject to the organization's oversight? This could be called the "perimeter" of the anti-corruption effort. It could be public administration narrowly defined, i.e. government employees only, or a subset of these, or it could include SOEs, or it could be a combination of these criteria.

1.3.3 The definition of corruption

The key element in a definition of corruption is whether the focus is the public or private sector. The UNCAC mostly concerns public administration and government operations but, in reality, government operations might also include SOEs. We might still speak of public corruption even when the private sector is involved to some degree: corruption in public administration often occurs in relation to its dealings with the private sector, such as in procurement practices.

A further element is the depth of the phenomenon that is implied by the term. A basic understanding of corruption is that it is a criminal activity, but it could also imply sub-criminal behaviour. As such, "corruption" could simply refer to inefficient or ineffective public administration. The appropriate term to use here is "maladministration".

1.3.4 Prevention vs. repression

Repression is characterized by an anti-corruption organization that focuses on or is specialized in a more judicial approach. Prevention means a focus on the removal of the causes and the organizational arrangements that may lead to opportunities for corruption. This item is related to the measures the anti-corruption organization either enacts itself or sets forth in its recommendations. Repression is mostly based on existing mandates, whereas prevention emphasizes managerial change in overseen organizations. The latter can be effected via general organizational measures, such as personnel rotation. Organizational measures are also linked to an area that has been developed in the private sector: risk management. Change can also take the form of measures to remove certain individuals and install replacements.

1.3.5 Macro vs. micro relationships in the field

Contact with "the field" is an important consideration. The basic dichotomy here is whether an anti-corruption organization entertains direct contact and reporting from the field, which we would call a "micro" approach, or whether it concentrates its operations on a few big issues. The meaning of "big" in this context has to be defined, of course, but the idea is clear: the latter type of organization leaves the overseen organizations to deal with micro-reporting from the field or else it leaves it to other overseen organizations—such as the judiciary—to deal with cases that it does not consider of interest in the fulfilment of its mission.

1.3.6 The focus of measures: objective vs. subjective

Whether through direct action or via recommendations, the types of measures an organization adopts also go towards defining its overall approach to anti-corruption. These can be of two different types. "Objective" measures are those geared to change the organization of overseen public administrations. With this term we are not encompassing the whole structure of a public entity, but we are referring to managerial systems, its strategy and its personnel management. As mentioned above, risk management is part of organizational management. Organization-specific corruption prevention plans are objective measures. "Subjective" measures, on the other hand, are designed to target the characteristics of individual members of the organization. For instance, they aim to ensure the impartiality of managers and to define and avoid conflicts of interest. This objective–subjective dichotomy is of most relevance to the prevention type described in point four above.

1.3.7 The role of transparency

Transparency is undeniably a companion of anti-corruption. Many sources attest to this. It should be made clear that an important part of tackling corruption through administrative measures and from the preventative perspective involves the enhancement of transparency mechanisms, adhering to the old adage that "sunlight is the best disinfectant" (Brandeis, 1914; Cordis and Warren, 2014).

Transparency measures operate on different levels, and with different aims (Birkinshaw, 2006; Heald, 2006; Merloni, 2008; Arena, 2008; Cerrillo Martinez, 2011). Above all, they guarantee the rights of citizens affected by administrative action, through the right of access to documented information on the activity of public administration and other institutions; thus, they essentially operate within the "due process" paradigm.

Its democratic and participatory dimension is also traditionally thought of as an important aspect of the role of transparency (Carloni, 2014).

The pursuit of transparency began independently of anti-corruption measures, with freedom of information legislation. Thus the promotion of transparency was formalized within public administrations before the advent of anti-corruption action. It is therefore interesting to note how the conceptual link between transparency and anti-corruption has been resolved within the framework of anti-corruption action and institutionalization.

This seven-dimensional framework for analysis is not, in fact, easy to implement in practice. The dimensions are far from mutually exclusive or collectively exhaustive. It may not be possible to stick rigidly to the framework; nonetheless, it represents a good starting point, and there will be time and space for further considerations and refinements. A framework is invaluable in helping us sift through institutional descriptions and in identifying specific meanings, which are not always overt or indeed the key objective of the communication. We will now proceed to apply this descriptive framework to our chosen set of countries.

1.4 Country case: France

The 2018 Corruption Perception Index (CPI) ranks France, regarding the level of corruption within its public sector, at 22 out of 176 countries, with a score of 71 out of 100 (Transparency International, 2018). Control of corruption in France was assessed as 89% (Transparency International, 2018).

France's anti-corruption organization is the Agence Française Anticorruption (AFA), a very young institution, with its mandate established in 2016.

1.4.1 The institutional setup

1.4.1.1 Independence from the branches of power

The AFA was established with Public Law no. 1691 of 9 December 2016, also called the "Sapin II" law after its promoter (Broussolle, 2017). This law reformed transparency and redefined the framework of anti-corruption in France. The new agency replaced the previous Service Central de Prévention de la Corruption (SCPC) and followed the Italian example: in its report for 2015 (*Rapport pour l'année 2015*), the SCPC cited the Italian Anti-corruption Authority as a model.

The agency is a comparatively small body with 70 staff, including experts from judicial and administrative backgrounds, who are known as agents. The annual budget is €10–15 million. The agency is headed by a magistrate, appointed by decree of the President of the Republic. It has national jurisdiction but is not independent from the executive government, being attached to the Ministry of Justice and periodically (once a year) reporting to the Prime Minister and to the Minister of Justice.

4.1.1.2 The shape of organizational arrangements

The agents play an important role in the AFA: authorized by decree of the Council of State, they can request and check any professional document, whatever the medium, or any information from the administrations and organizations under the agency's control. Moreover, they can proceed uninvited on-site in pursuit of verification of the accuracy of the information provided by the public administration organizations, or may interview, in conditions ensuring the confidentiality of their exchanges, any person whose assistance they deem necessary.

1.4.1.3 Modus operandi

The decree of 14 March 2017, no. 2017–329, defines the AFA's structure and organization, stipulating a consulting department ("conseil stratégique") and a control department ("commission des sanctions") for public actors. The consulting department develops and updates recommendations to prevent and detect offences: corruption, trading in influence, extortion, illegal capture of interest, misappropriation of public funds and favouritism. In this endeavour, the agency must prepare a multi-year plan. The control department controls the quality and the efficiency of the procedures implemented in the administration of the state, in the local authorities, in the public institutions, in the SOEs, the associations and the foundations that are recognized to be of public utility.

In addition, the agency's powers of inspection include the right to demand the formulation and implementation of anti-corruption programmes (or "compliance programmes") from the organizations it oversees, i.e. SOEs and public administration.

The agency's key functions involve centralizing and analysing corruption data, as well as providing advice on corruption-related matters to other authorities, including judicial authorities, and public institutions. The agency analyses corruption at a systemic level country-wide in order to better understand how corruption evolves. It provides recommendations on corruption prevention which it presents in an annual report.

1.4.2 The power and scope of the organization

The AFA began its anti-corruption work in those SOEs with over 500 employees and with a turnover of over €100 million. Activities within public administration began in 2018. The agency has national jurisdiction.

1.4.3 Public vs. private

The principal jurisdiction of the AFA is over SOEs. In that sense, the agency only has authority over public corruption, albeit focusing on activities of a private nature. As noted above, in 2018 its scope broadened to include all public administration.

1.4.4 Mission: prevention vs. repression

The AFA's mission is preventative, with the aim of creating "collective discipline" vis-à-vis the risk of corruption. Thus, it supervises, advises, guides and regulates SOEs and public administration organizations in areas of corruption, trading in influence, extortion, illegal capture of interest, misappropriation of public funds and favouritism—a complex task formerly carried out by the discontinued SCPC.

In the public procurement domain, the Sapin II law grants the agency power to prevent corruption and direct public administrations towards correct tendering processes.

The agency does not pursue judicial cases unilaterally: it works in cooperation with law enforcement organizations such as the Office Central de Lutte contre la Corruption et les Infractions Financières et Fiscales (OCLCIFF), which is part of the Ministry of the Interior with authority to embark on any investigative action regarding corruption and tax fraud. The

AFA also develops joint programmes with public and private companies to provide corruption-prevention advice by means of improving their codes of ethics or providing training.

The agency is required to centralize and disseminate information necessary for the detection and prevention of both passive and active corruption offences and of breaches of probity. It offers training, awareness and assistance to state administrations and local authorities.

1.4.5 Operations: (micro-)reporting from the field or concentration on big issues

The AFA began its activities in 2017 by taking charge of "six big notifications" about large companies that required a step-by-step approach. The initial review was followed by on-site visits by the agent teams. As stated, its jurisdiction over other public entities only began in 2018. In the words of a prominent government officer, the main scope of the agency is to try to

> discipline the structures collectively [...] [T]here will always be someone looking for an envelope; no legislation can prevent that kind of phenomenon, but by disciplining the structures collectively, telling them to put safeguards in place, it makes things more difficult.
>
> (*France Soir*, 2017, author's translation)

1.4.6 The focus of action: objective measures or subjective impartiality and conflict of interest

The agency appears to be more concerned with objective organizational measures than subjective ones, although codes of ethics are part of its agenda. The "safeguards" referred to by the director in the interview just quoted may include subjective measures.

1.4.7 The relationship with transparency

In the field of transparency, the last paragraph of Section 1.4.4 is relevant here. However, the agency has no actual authority in matters of transparency. This sector of public policy is governed by the Haute Autorité pour la Transparence de la Vie Publique (Public Law no. 2013–907 of 11 October 2013), a fully fledged independent authority with a complex brief: to "spontaneously" reinforce the integrity of elected officials, and to combat corruption and modernize economic life. However, the relationship between the two bodies is regulated by the Sapin II law.

1.5 Country case: Italy

Italy scored 47 out of 100 on the Corruption Perception Index 2018—with zero being highly corrupt and 100 very clean (Transparency International, 2018)—a score that ranks Italy 60th out of 176 countries worldwide, and 26th out of 28 in Europe. In 2010, Italy came 67th out of 178 countries. The 2017 Index of Economic Freedom scored Italy's government integrity as 40.1, ranking it at number 79 out of 186 countries worldwide and 36th out of 45 in the European region (Miller, Kim and Robert, 2018). This same study concluded that

> Italy's economy, the Eurozone's third largest, is hobbled by exceptionally high public debt and such structural impediments to growth as labour market inefficiencies, a sluggish judicial system, and a weak banking sector. Political uncertainty increases the potential for financial volatility and could further delay structural reform. The economy remains burdened by political interference, corruption, and poor management of public finance. The complexity of the regulatory framework and high cost of conducting business cause considerable economic activity to remain in the informal sector.
>
> (Miller *et al.*, 2018)

Italy's anti-corruption organization is the Autorità Nazionale Anticorruzione (ANAC), the mandate of which began in 2014.

1.5.1 The institutional setup

1.5.1.1 Independence from the branches of power

ANAC operates with independence but not in isolation, because it reports to parliament and is functionally linked to the executive, specifically to the Department of Public Administration. In fact, its authority is appointed by the executive and approved by parliament. It has a staff of 300 and an annual budget of about €60 million.

1.5.1.2 The shape of organizational arrangements

The Authority has a functional field link in each government organization: the Corruption Prevention Officer, the role of which is as a facilitator but whose office will not necessarily have any staff. There are around 11,000 Corruption Prevention Officers in the Italian public administration, most located in municipalities, of which the country has over 8,000. Its shape is that of a central agency with a functional (non-hierarchical) relationship to officials within the supervised agencies. A Corruption Prevention Officer's responsibilities also include transparency.

1.5.1.3 Modus operandi

The Authority requires public administration organizations to draft an annual update to the Three-Year Plan for Transparency and Prevention of Corruption (PTPCT). It then runs a planning and feedback cycle. The Authority also checks that every organization of public administration publishes their own specific code of ethics.

The Authority performs oversight on contracts for the procurement of public works. It also delivers an annual report to parliament, focused on the explication of indicators and preventative measures taken. In addition, it is tasked with producing a national anti-corruption plan.

1.5.2 The power and scope of the organization

The Authority oversees the wider public sector, including public administration and SOEs, a total of about 4 million public employees. Oversight does not include employees of the judiciary or of parliament but it does include public employees of local and regional government.

1.5.3 Public vs. private

The Authority's focus is on corruption within public organizations. Public organizations include SOEs.

It is interesting to note that the Authority has adopted a very broad definition of corruption, which includes "maladministration", i.e. ineffective government and misconduct "below the line" of criminal activity. This will also have an impact on the nature of the Authority's focus as regards the prevention–repression dichotomy, which must necessarily veer towards prevention, repression only being concerned with criminal activity.

1.5.4 Mission: prevention vs. repression

ANAC's emphasis is on prevention, although collaboration and reporting to the judiciary are part of the Authority's remit. Prevention takes the form of objective measures, such as the drafting of corruption prevention plans on the part of each individual organization within public administration. The Authority also scrutinizes public contracts, undertakes specific investigations and offers support to individual organizations. Subjective measures are proposed by the Authority as part of general legislation, and are also intended to be part of the plans of individual organizations.

1.5.5 Operations: (micro-)reporting from the field or concentration on big issues

ANAC undertakes specific operations on specific issues. However, its mandate also includes receipt of micro-reporting from the field, both from individuals (public employees or citizens) and from public or private organizations. It is possible that the orientation is beginning to change towards a concentration on large issues and large contracts.

1.5.6 The focus of action: objective measures or subjective impartiality and conflict of interest

ANAC pursues both objective and subjective measures. Prevention, as mentioned previously, takes the form of objective measures, such as the drafting of corruption prevention plans. Subjective measures seek to uphold the integrity of public officials mainly through removal of conflict of interest. The Authority also has a specific role in the procurement procedures of individual public administration organizations. In fact, oversight of (micro-)activity on the part of the Authority has an increasing impact on the structure and organization of the Authority itself. The Authority is also concerned with transparency of public procurement.

1.5.7 The relationship with transparency

Freedom of information was only fully enacted in Italy in 2016, following previous similar provisions dating back to the 1990s. Transparency was subsequently added to ANAC's mission, being recognized as an important element of anti-corruption action. However, some of the transparency provisions are outside the authority's purview. Transparency of public administration in Italy is pursued through two instruments: (1) a mandate for individual public administrations to publish specific data and information on the website of each administration; (2) the Freedom of Information Act, i.e. the generalized right of citizens to access information not published on the administrations' websites. The Authority is only responsible for the enforcement of the former, not the latter.

1.6 Country case: the United Kingdom

The Transparency International Corruption Perception Index places the UK in the top ten least corrupt countries with a score of 82 out of 100. The country ranks eighth out of 176 (Transparency International, 2018). The data also establishes that UK control of corruption is 97%. The data is corroborated by other perception studies including the 2018 Index of Economic Freedom which scored the UK's freedom from corruption and its government integrity at 79, ranking it fourth out of 45 countries in the European region and eighth out of the total of 186 (Miller, Kim and Robert, 2018).

1.6.1 The institutional setup

1.6.1.1 Independence from the branches of power

The Serious Fraud Office (SFO) is the key agency investigating and prosecuting cases of corruption, although other national, regional and local authorities (including the Metropolitan Police) have authority to deal with corruption-related offences. The director of the SFO reports to the Prosecutor General, who, in turn, is appointed by the Prime Minister and is responsible to parliament for the SFO. The SFO has a staff of 510.

On the executive side of anti-corruption efforts, every year the Prime Minister selects an Anti-corruption Champion to support the government in implementing its own anti-corruption strategy, a post that is also engaged with organized crime and large economic crime. The Anti-corruption Champion is supported by the Joint Anti-corruption Unit, an office located within the Home Office, i.e. the ministry, or department, for domestic security. Such a structural position is to ensure better coordination of domestic and international anti-corruption efforts and promote stronger links between anti-corruption and other units working on economic and organized crime.

The institutional picture in the UK appears to differ from the ones described so far. In this case the organizational body is positioned between the executive and the judicial branches. We find here a diversity of organizations, with no doubt overlapping roles and responsibilities, and a de facto emphasis on repression.

1.6.1.2 The shape of organizational arrangements

The organizational arrangement appears to be composed of an executive unit, the Joint Anti-corruption Unit, which is meant to "encourage" all other units concerned with crime, which are located in all the branches of government, not only in the judiciary. In fact, parliament also enters the picture, with a Committee on Standards in Public Life and a Parliamentary Commissioner for Standards, the Adviser on Ministerial Interests, and the Advisory Committee on Business Appointments. The overall organizational shape appears to be one of multiplicity: a network rather than a pyramid. UK anti-corruption strategy appears to have no central body with the role of an anti-corruption authority.

1.6.1.3 Modus operandi

To create an actual network of different bodies, the Joint Anti-corruption Unit lists four ways of working that can be summarized as follows: "strong leadership", identification of risks (including insider threat), training and education, and repression (which appears to be the prevailing option). Criticism is sometimes levelled to the effect that this comes across merely as a statement of intent, rather than as a range of active programmes.

Interesting among the objectives is that of redress from injustice as a result of corruption, whereas a standard provision is that of support to those who report corruption (whistleblower protection).

Activities of bodies other than the Joint Anti-corruption Unit revolve around conflict of interest. For instance, an Advisory Committee on Business Appointments, "sponsored" by the Cabinet Office, advises on "pantouflage", i.e. appointments for ex-politicians and high-echelon public employees.

Regarding inside-out reporting—i.e. anti-corruption organizations reporting about their own anti-corruption activity—as mentioned, the SFO delivers an annual report to the Prosecutor General, whereas the Joint Anti-corruption Unit writes a document on national anti-corruption strategy. This document outlines a taxonomy of its own concerns under six headings: insider threat, public- and private-sector integrity, global business environment, public procurement, collaboration with other countries, and corruption in finance.

1.6.2 The power and scope of the organization

Both the Joint Anti-corruption Unit and the SFO have nationwide authority, apparently at all levels of government, an authority that is exercised in both public and private matters. Emphasis appears to be placed on private businesses, since international activities are specifically mentioned. The focus is on white-collar crime.

1.6.3 Public vs. private

The SFO has a mandate to investigate in cases of both public and private corruption. International corruption, i.e. corruption in private businesses dealing with foreign governments (foreign corrupt practices), receives more attention than domestic corruption, the latter sometimes downplayed as misuse of funds or other minor offences. There appears to be no major emphasis on corruption in public administration.

1.6.4 Mission: prevention vs. repression

The Joint Anti-corruption Unit appears to be pursuing a strategy of broadening the anti-corruption vision from one of judicial action. Its "strategy against corruption" is attempting to move beyond judicial action, without impairing judicial action. This could be described as prevention, while the judicial units—including the SFO—are concerned with repression. The focus appears to be on foreign corrupt practices.

1.6.5 Operations: (micro-)reporting from the field or concentration on big issues

A wide variety of channels exists for citizens to report suspected corruption, but the Anti-corruption Champion and the Joint Anti-corruption Unit are not among them as the Joint Anti-corruption Unit has no relationship with the public. Citizens can report to a number of other bodies, like the police or the judiciary.

Likewise, the SFO concentrates only on "serious or complex fraud" and chooses which cases to take on. Cases are reported to the SFO by different sources: law enforcement agencies, regulators, other authorities, whistleblowers and self-reporting from corporate entities. Cases mostly concern white-collar crime in the business and financial sector. The SFO accepts 15–25 new cases each year. For example, in 2016–2017 a total of 13 defendants were convicted in seven cases, giving conviction rates by defendant of 86.7% and by case of 100%. This is an interesting measure of effectiveness, which does not appear to be available in other countries. We conclude that anti-corruption institutions in the UK concentrate on big issues and do not entertain a relationship with the field, although the SFO does receive reports from a wide variety of sources, including—apparently—the general public.

1.6.6 The focus of action: objective measures or subjective impartiality and conflict of interest

Bodies of a judicial type—like the SFO—necessarily work on subjective measures of individual responsibility: crime, ultimately, is committed by individuals. More subjective measures are available in the Civil Service Code and the Seven Principles of Public Life drawn up and supervised by the parliamentary Committee on Standards in Public Life. The seven principles listed are of interest: selflessness, integrity, objectivity, accountability, openness, honesty and leadership.

Objective measures—of an organizational nature—are the specific concern of the Joint Anti-corruption Unit. As previously mentioned, we could include here identification of risks (including insider threat) and general measures such as training and education.

1.6.7 The relationship with transparency

Transparency occupies a central position here, because the UK has defined an "ecosystem" of different mechanisms (Kreimer, 2008; Richards and Smith, 2015): "A mixture of openness instruments, technology and accountability have created a system where freedom of information laws join more dynamic instruments, from social media to mass leaks" (Birchall, 2014, p. 84; Kreimer, 2008). The UK is part of the Open Government Partnership. Transparency is one of the concerns of the Joint Anti-corruption Unit, yet there is no evidence of an institutional link between tasks related to transparency and those of anti-corruption.

1.7 Country case: the United States of America

In 2016, the Corruption Perception Index ranked the level of corruption within US public sector as 18th out of 176 countries, with a score of 74 out of 100 (Transparency International, 2018), three points above France.

1.7.1 The institutional setup

1.7.1.1 Independence from the branches of power

The US Office of Government Ethics represents the key component of that country's approach to anti-corruption. The appointment of the head of the Office of Government Ethics is made by the President and is subject to congressional approval. The appointee serves for only one mandate and the office does not present a formal, hierarchical report to the President.

The Office of Government Ethics was established through the Office of Government Ethics Act in 1978, well before the UNCAC. Germane contemporary legislation was the Inspectors General Act of 1978, which established the complementary institutional figure of the Inspector General within several executive agencies.

The Office of Government Ethics has a small staff, a small budget and no field personnel. However, it is the central organization acting on behalf of the Inspectors General, who are appointed in each government agency. Each Inspector General is also subject to congressional approval. The Inspectors General are investigators, as well as agency promoters of an anti-corruption climate.

This anti-corruption legislation followed on the heels of major corruption scandals in private business involving the bribing of foreign officials (e.g. the bribery scandal involving the aircraft manufacturer Lockheed). It was also preceded by significant domestic scandals, such as Watergate. This legislation, then, was forged in pursuit of both public anti-corruption and anti-corruption in foreign practices.

1.7.1.2 The shape of organizational arrangements

The Office of Government Ethics is a central agency with a functional (non-hierarchical) relationship to officials within the supervised agencies.

1.7.1.3 Modus operandi

The Office of Government Ethics provides direct support—through desk specialists—to the 4,500 officials in the Inspectors General offices (or "agency ethics officers" as per the document *Standards of Ethical Conduct for Employees of the Executive Branch*). The Office of Government Ethics also undertakes training, "institution building", "aggregating" and support activities. It prepares training materials through its education division; subjects dealt with include: the misuse of position, gifts between employees, working with contractors, and gifts from outside sources (United States Office of Government Ethics, *Preventing Conflicts of Interest*, 2016).

1.7.2 The power and scope of the organization

The mandate of the Office of Government Ethics is to supervise the 2.7 million civilian employees of the executive branch of the federal government. The total count of public administration employees in the USA is about 24 million. Neither the Office nor the legislation establishing anti-corruption action are concerned with state and local public administration. The Office supervises no branches of power other than the executive branch. Public–private partnerships also appear to be outside its scope.

1.7.3 Public vs. private

The Office of Government Ethics is concerned only with corruption in the US public administration. Private corruption is left to the Foreign Corrupt Practices Act, which is enforced by the judiciary and mostly concerns US corporations interacting with foreign governments—so, although this is technically about public administrations (foreign ones), it remains outside the Office's domain.

1.7.4 Mission: prevention vs. repression

The Office of Government Ethics is concerned with both repression and prevention. On the repression side, the Inspectors General, with whom the Office interacts, conduct investigations and refer cases to the judiciary.

1.7.5 Operations: (micro-)reporting from the field or concentration on big issues

The Office of Government Ethics does not interact with the field, i.e. citizens or individual employees of public administration; it interacts only with the offices of the Inspectors General.

1.7.6 The focus of action: objective measures or subjective impartiality and conflict of interest

The Office of Government Ethics is responsible for issuing regulations on conflicts of interest and publishes an annual report about cases referred to the judiciary and their outcomes: the Survey of Prosecutions. It also publishes the *Standards of Ethical Conduct for Employees of the Executive Branch*. The emphasis therefore appears to be on subjective rather than objective measures of an organizational type.

1.7.7 The relationship with transparency

Transparency is a concern of Office of Government Ethics activities, although implementation of the Freedom of Information Act rests with the judiciary.

1.8 Country case: Australia

Australia scored 85 out of 100 on the 2016 Corruption Perception Index (Transparency International, 2018), which ranks it as seventh out of 176 countries worldwide. In 2010, it scored 96 and was identified as one of the least corrupt countries in the world (Transparency International, 2018). Other perception studies include the 2013 Index of Economic Freedom which scored Australia's freedom from corruption as 88, ranking it eighth out of 42 countries in the Asia Pacific region (Miller, Kim and Robert, 2018).

1.8.1 The institutional setup

1.8.1.1 Independence from the branches of power

Australia, like the US, is a federation of states. There is no central body with anti-corruption functions at federal level but there are anti-corruption institutions in the individual states of New South Wales, South Australia and Victoria. The state of Queensland has a Crime and Corruption Commission. There is no independent anti-corruption body in the Northern Territory.

A generic model of an anti-corruption institution is represented by the Independent Commission Against Corruption. The discussion here is mostly based on the New South Wales Independent Commission Against Corruption (ICAC), which is independent of the branches of power and has been awarded special powers to perform its investigative functions.

1.8.1.2 The shape of organizational arrangements

Some researchers have identified a noticeable gap in the oversight of the Australian public sector which leaves several institutions unsupervised, such as the federal parliament and federal agencies (Australian Collaboration, 2013). The only relevant federal institution, the Commonwealth Ombudsman's Office, does not conduct investigations.

1.8.1.3 Modus operandi

The New South Wales ICAC is modelled on the Hong Kong Independent Commission Against Corruption, one of the "most replicated and widely recognised examples of best practice in anti-corruption, globally" (Quah, 2009). The New South Wales body was established in 1989.

Queensland's Crime and Corruption Commission has a witness protection programme.

1.8.2 The power and scope of the organization

The New South Wales ICAC investigates corruption in the state's public sector, including SOEs. The public sector under its oversight has approximately 320,000 employees.

1.8.3 Public vs. private

The role of the ICAC is focused on the public sector and cases of public corruption.

1.8.4 Mission: prevention vs. repression

The functions of the ICAC include provision of advice and guidance via information, resources and training to public-sector agencies. The Commission also conducts research to identify corruption risk. In the year 2016–2017, it issued 105 separate advisory documents. The Commission combines different powers, with its most important function being investigations, i.e. repression of corruption. Nonetheless, prevention is not ignored: the objectives in the *Strategic Plan 2017–2021* (ICAC, 2017, December) for preventing corruption encourage the government to address corruption risks.

1.8.5 Operations: (micro-)reporting from the field or concentration on big issues

The ICAC receives and analyses complaints from members of the public and public officials, and reports made by the principal officers of public-sector agencies and ministers of the Crown. As benchmarks of the organization's success, its 2016–2017 Annual Report cited the following: matters received or referred to the organization: 2,489; full investigations: 10. It states that its corruption prevention initiatives included the delivery of 74 training sessions and that 89% of corruption-prevention recommendations were implemented (NSW ICAC, 2017). It can be concluded that the Commission does engage in micro-reporting but at the same time focuses its attention on a few big issues. As such, in encompassing micro-reporting along with big issues, a large amount of work is implied in selecting only what is relevant.

1.8.6 The focus of action: objective measures or subjective impartiality and conflict of interest

The Commission's measures are mostly subjective, performing as it does investigations and judicial prosecutions. However, it also promotes objective measures when it makes change recommendations to the executive.

1.8.7 The relationship with transparency

Freedom of information and transparency legislation was introduced at a federal level in Australia in 1982 and therefore the states' anti-corruption authorities are not concerned with this issue. In 2010, freedom of information administrative and policy functions were centralized into an independent agency, the Office of the Australian Information Commissioner. However, freedom of information legislation was also introduced within each state and territory, but this does not appear to be related to anti-corruption efforts.

1.9 Summary

The country cases set out in this chapter have allowed us to glimpse the reality of anti-corruption efforts. At first glance, it may seem that the approaches among these diverse countries are all much the same. There is a sense that the descriptions offered—according to the seven identified elements—do not go deep enough. But getting to the heart of what these anti-corruption organizations are all about is far from straightforward: we need to dive beneath the surface of the statements of intent. We also know from experience that things that share the same name are not the same from country to country. There are differences in absolute levels of performance and in efficiencies of execution. We need to dig deeper to find out what is meant by phrases like "risk avoidance". What is a subjective measure, exactly, and how is a corruption prevention plan written? We need to explore how these functions are actually pursued. And we need numerical data with which to appreciate the scope of anti-corruption measures.

Our initial plunge into the world of anti-corruption has produced more questions than answers. The descriptions offered in this chapter are clearly insufficient on their own: breadth should not be at the expense of depth. We need now to continue with an implementation approach, which requires a more detailed study of anti-corruption practice. To this end the next two chapters will consist of an in-depth presentation of one specific anti-corruption organization in one specific country: Italy.

As shown above, in terms of corruption, Italy ranks lowest among the five countries observed here. This in itself is a strong reason to examine what it is doing with regard to anti-corruption, with a view to stimulating criticism and evaluating its position. Chapter 2, "The mission and operations of the Italian experiment", will examine the various steps that have been taken to establish the Italian National Anti-corruption Authority, and Chapter 3, "An inside-out view of Italy's Anti-corruption Authority", will present in detail the organizational structure and the individual functions of the Authority's bureaux.

Note

1 Kofi A. Annan, Secretary General of the United Nations, UN Office on Drugs and Crime, Vienna; UN Convention Against Corruption, New York, 2004.

Chapter 2

The mission and operations of the Italian experiment

2.1 Introduction

Italy's anti-corruption effort began in 2009 and was relaunched in 2014 when Prime Minister Matteo Renzi came to power. The new cabinet, determined to be more assertive, established the National Anti-corruption Authority (ANAC) through Law Decree no. 90 of 2014, converted with modification into Public Law no. 114 of 2014. The basic idea was to transform the Authority from a mere think-tank into an independent authority. It is not dependent on the government, i.e. the cabinet, but it reports to parliament. The Authority is headed by Raffaele Cantone, a judge who, in his role as public prosecutor, had previously presided over a series of successful trials against the mafia.

The operating concept of the Authority describes a universal approach, across the ambit of public administration and the public sector, with a wide range of measures at its disposal, addressing several aspects of public administration organization and systems from personnel to contract law. Its role is one of a mentor to each organization's own, self-managed anti-corruption activities. It is also charged with a complex set of tasks, including supervision of public procurement. The onus, in fact, is placed on each individual organization to tackle corruption itself, with the Authority seen simply as a facilitator or possibly a guardian. The Authority was once jokingly referred to as a "big policeman", with a mandate to put everybody into jail. The Authority rejects this notion and in fact labours in quite the opposite direction.

The range—or organizational perimeter—of Authority supervision covers not only organizations that are fully part of public administration (3.3 million employees) but also those that are owned by or have a significant (e.g. golden share) participation in public administration (Cantone and Merloni, 2015).

What follows is a timeline summary of events and documentation produced within the Authority's corruption-prevention cycle. We will subsequently discuss each event and document in detail, adding relevant information along with the Authority's own perspective.

The following two sections therefore represent our effort to understand the specifics and evolution of Anti-corruption Authority action, to describe those specifics within the framework of ANAC activities, and (not least) to translate into English the legal and organizational terminology.

2.2 Timeline of Anti-corruption Authority key statutes and events

15 February 1999
OECD Convention on Combating Bribery of Foreign Public Officials in International Business Transactions.

30 November 2000
Freedom of Information Law is adopted in UK.

9 December 2003
United Nations Convention Against Corruption (UNCAC or Merida Convention) is signed. Article 6 of the Convention requires signatory countries to set up an anti-corruption authority.

2 July 2009
GRECO, the European Council's Group of States against Corruption, begins its reviews of Italy—joint first and second evaluation round, evaluation report on Italy. First evaluation round: independence, specialization and means available to national bodies engaged in the prevention of and fight against corruption; extent and scope of immunities. Second evaluation round: proceeds of corruption; public administration and corruption; legal persons and corruption.

27 October 2009
Establishment of an "evaluation commission" and a Transparency Portal project (Legislative Decree no. 150 of 2009). The Commission was named the Independent Commission for Evaluation, Transparency and Integrity, its acronym being CIVIT.

17 November 2009
First Report to parliament of the Anti-corruption and Transparency Service (SAeT) which was part of the Ministry for Public Administration.

19 January 2010
First bill ("draft law") by the Minister of Justice of the Berlusconi Cabinet, Angelino Alfano (the "Alfano" draft law of 2010).

6 May 2011
Second Report to parliament of SAeT.

23–27 May 2011
GRECO: compliance report on Italy on the joint first and second evaluation round.

20–23 March 2012
GRECO: third evaluation round, evaluation report on Italy. Theme I: incriminations. Theme II: transparency of party funding. Compliance report on Italy, 16–20 June 2014.

1 October 2012
The cabinet, led by Prime Minister Monti, approves a report by the Commission for the Study and Elaboration of Proposals on Transparency and Prevention of Corruption in Public Sector. It includes amendments on the Berlusconi cabinet draft law of 2010.

6 November 2012
The Independent Commission for Evaluation, Transparency and Integrity is made the National Anti-corruption Authority under Public Law no. 190 of 2012. This law emphasizes corruption-prevention measures.

17–21 June 2013
GRECO: addendum to the second-round compliance report on Italy.

31 August 2013
Legislative changes are approved on the Anti-corruption Authority's mandate: Public Law no. 125 of 2013.

11 September 2013
The Anti-corruption Authority approves the first National Anti-corruption Plan (PNA), including specific coordination measures for public administration organizations to write their Three-Year Plans to Prevent Corruption, which were mandated by Public Law no. 190 of 2012.

20 September 2013
OECD Integrity Review of Italy: *Reinforcing Public Sector Integrity, Restoring Trust for Sustainable Growth*.

30 October 2013
Conversion of Law Decree no. 101 of 2013 into Public Law no. 125 of 2013.

18 December 2013
Report on the first year of implementation of Public Law no. 190 of 2012 which established the National Anti-corruption Authority.

31 January 2014
Deadline for the first approval and publication of the first Three-Year Plan to Prevent Corruption, as mandated by Public Law no. 190 of 2012.

4 April 2014
Appointment of Raffaele Cantone as President of the "new" Anti-corruption Authority by decree of the President of the Republic.

16–20 June 2014
GRECO: compliance report on Italy, third evaluation round, evaluation report on Italy. Theme I: Incriminations; Theme II: Transparency of party funding.

24 June 2014
The "new" Anti-corruption Authority is established through Law Decree no. 90 of 2014.

11 July 2014
Appointment of four new Council Members of the Anti-corruption Authority. The Council of the Anti-corruption Authority has five members in all, the fifth being the President.

11 August 2014
Law Decree no. 90 of 2014 is converted into Public Law no. 114 of 2014.

18 November 2014
The Authority issues Resolution no. 146 of 2014 on "power of jurisdiction", which concerns the exercise of power of order by the Authority in the case of non-adoption of acts or measures required by the National Anti-corruption Plan or the Three-Year Plan to Prevent Corruption, or by the rules on the transparency of administration, or in the case of conduct contrary to the plans and the rules on transparency.

31 January 2015
Deadline for publication of the Three-Year Plan to Prevent Corruption for 2015–2017 as mandated by Law no. 190 of 2012.

1 February 2015
Evaluation of the Three-Year Plan for 2015–2017 by the Anti-corruption Authority begins, as mandated by Law no. 190 of 2012.

2 July 2015
The Anti-corruption Authority publishes its Annual Report to parliament for the year 2014.

14 July 2015
The Anti-corruption Authority holds its first national meeting with the individual public administration and wider public-sector organizations' Corruption Prevention Officers.

28 October 2015
The Anti-corruption Authority approves its 2015 Update to the National Anti-corruption Plan 2013–2015.

31 December 2015
Corruption Prevention Officers draft their first report on the implementation of the Three-Year Plan to Prevent Corruption for 2015–2017.

31 January 2016
Deadline for publication of the Three-Year Plan for 2016–2018 by the individual public administration and wider public-sector organizations, as mandated by Public Law no. 190 of 2012.

1 February 2016
Evaluation of the Three-Year Plan to Prevent Corruption for 2016–2018 of the individual public administration and wider public-sector organizations begins on the part of the Anti-corruption Authority.

The Anti-corruption Authority approves its own Reorganization Plan after the merger between the Public Procurement Authority (AVCP) and CIVIT.

19 April 2016
The new Public Procurement and Concession Contract Code is adopted through Legislative Decree no. 50 of 2016.

24 May 2016
The Anti-corruption Authority holds its second national meeting with the individual public administration and wider public-sector organizations' Corruption Prevention Officers.

25 May 2016
The government approves the review and simplification of the provisions in the field of prevention of corruption, publicity and transparency through Legislative Decree no. 97 of 2016 (on freedom of information).

22 June 2016
First Whistleblower Monitoring Report.

14 July 2016
The Authority's *Annual Report 2015* to parliament.

3 August 2016
The Authority approves the new National Anti-corruption Plan for 2016–2018.

24 April 2017
Law Decree no. 50 art. 52 grants the Anti-corruption Authority self-regulation on organization and personnel, following the model of independent administrative authorities.

24 May 2017
The Anti-corruption Authority holds its third national meeting with the individual public administrations and wider public-sector organizations' Corruption Prevention Officers.

6 July 2017
The Authority's *Annual Report* on 2016 activities to parliament.

22 November 2017
The Anti-corruption Authority approves its 2017 Update to the National Anti-corruption Plan (PNA) 2016–2018.

30 November 2017
Parliament approves Public Law no. 179 of 2017 on whistleblower reform.

2.3 Towards a cycle of planning and feedback

In this section we elaborate on some of the key events on the timeline, illustrating the evolution of the anti-corruption effort.

27 October 2009
Establishment of an Independent Commission for Evaluation, Integrity and Transparency (CIVIT) and of the Portal of Transparency project—Legislative Decree no. 150 of 2009.
The cabinet, led by Prime Minister Berlusconi, establishes the Commission for Evaluation, Integrity and Transparency in Organizations of Public Administration (CIVIT). It is established and governed by Legislative Decree no. 150 of 2009 (art. 13), to oversee the implementation of evaluation and performance measurement of public administration managers and employees. This function is to be performed by individual public administration organizations through their Performance Plans. The Commission is considered a quasi-independent bureau of the Department of Public Service, which is part of the executive branch, and it is staffed with civil service personnel. Notwithstanding the fact that the legislative order mentions the integrity of public

employees, the Commission is not explicitly tasked with corruption in public administrations.

Legislative Decree no. 150 of 2009 also launched the Portal of Transparency project, with the aim of making the activities of public administrations known and accessible. The portal is intended to make available information about strategic missions, indicators and objectives of central government, and on the quality of public services. The legislative decree also established the creation in each public organization of an Independent Evaluation Unit (OIV) (Di Mascio and Natalini, 2013). Finally, the legislative decree established that each public organization should write and publish a Three-Year Transparency Programme.

19 January 2010
First draft public law by the Minister of Justice.

6 November 2012
CIVIT Public Law no. 190 of 2012.
The cabinet, led by Prime Minister Monti, proposes Public Law no. 190 of 2012. Article 1, paragraph 2 identifies CIVIT as the National Anti-corruption Authority. This Public Law fulfils international obligations that require Italy to set up an authority for the identification of political corruption and the adoption of preventative measures to tackle corruption in the public sector. CIVIT was relatively independent from the Department of Public Service; however, its new anti-corruption mandate overlaps with the remit of the Department of Public Service. For instance, the National Anti-corruption Plan (PNA) (mandated by the same Public Law) was approved by CIVIT but drafted by the Department of Public Service.

31 August 2013
Legislative changes make CIVIT an anti-corruption authority (Public Law no. 125 of 2013).
The cabinet is now led by Prime Minister Letta. CIVIT undergoes a further legislative amendment by Law Decree no. 101 of 2013, subsequently undergoing significant change upon conversion into Public Law no. 125 of 2013, at which point the members of the Commission are replaced as well. The procedure is confirmed for the appointment of the three members and a clear distinction is made: on the one hand, functions in the area of performance appraisal are entrusted to the Negotiating Agency for Public Administrations; on the other hand, functions of anti-corruption and transparency are entrusted to CIVIT. CIVIT adopts the official name of the National Anti-corruption Authority and for the Evaluation and Transparency of Public Administration (ANAC). However, difficulties persist regarding lack of clarity about allocation of tasks between the Authority and the Department of Public Service.

11 September 2013
The National Anti-corruption Plan 2014–2016.
The National Anti-corruption Authority (ANAC) is approved, acting on a proposal of the Department of Public Service, namely the National Anti-corruption Plan (PNA), in accordance with Article 1, Paragraph 2, Item b of Public Law no. 190 of 2012. The plan is drawn up based on the guidelines of an inter-ministerial committee. It contains the government's strategic object-ives for the development of a centralized prevention strategy and provides guidance and support to public administrations for the implementation of corruption prevention and for the drafting of each public administration's own Three-Year Plan to Prevent Corruption.

This National Plan (2014–2016) implements Public Law no. 190 of 2012, in which it is called "Measures for the prevention and repression of corrup-tion and illegality in the public administration", or the National Anti-corruption Plan (PNA, 2013 [24 July]).

30 October 2013
Conversion of Law Decree no. 101 into Public Law no. 125 of 2013.
Law Decree no. 101 of 2013 is converted in Public Law no. 125 of 2013 and undergoes many changes. This conversion process restores to CIVIT/ANAC all the functions relating to performance evaluation, entrusted tem-porarily to the Negotiating Agency for Public Administrations (ARAN), including supervision over individual public administration organizations' Performance Plans via those organizations' Independent Evaluation Unit (OIV).

The law also reformulates the composition of the Authority's governing body, as follows:

> a collegial authority composed by a chairman and four members chosen among highly professional experts, including those external to public administration, with proven expertise in Italy and abroad, in both the public and the private sector, of recognized independence and proven experience in the field of containment of corruption, of management and of performance measurement.

The independence of the members is guaranteed by a requirement of non-provenance from political or trade union office in the three years before appointment and is reinforced by the prohibition of a second term—a stipu-lation that is also amply justified by the length of the term of office, namely six years. Regarding the proposal to the Council of Ministers for the appoint-ment of the four members, the Minister for Public Administration independ-ently opted for an open procedure for the call for applications, CVs and outlines of the activities to implement, with a view to further enhancing the credibility and independence of the nominees.

31 January 2014
Deadline for publication of the Three-Year Plan to Prevent Corruption 2014–2016.
The individual organizational units of public administration (and the public sector) are required to publish their Three-Year Plan to Prevent Corruption on their own websites, in a specific area named "Transparent Administration", which should contain easily accessible information on the most important details concerning institutional bodies, executives, managers and activities undertaken.

4 April 2014
Appointment of the President.
The Renzi Cabinet appoints Raffaele Cantone as President of the "National Anti-corruption Authority and for the Evaluation and Transparency of Public Administrations", by decree of the President of the Republic.

24 June 2014
The Commission (CIVIT) becomes the Anti-corruption Authority (ANAC).
Law Decree no. 90 of 24 June 2014, converted into Public Law no. 114 of 2014, renamed the CIVIT/ANAC Commission the National Anti-corruption Authority (ANAC) and transferred the functions of measurement and evaluation of performance back to the Department of Public Service, which is governed by the Minister for Public Administration. The decree brings within the Authority structure the former Authority for the Supervision of Public Contracts (AVCP) in order to afford the Anti-corruption Authority sufficient organizational and economic resources. The decree grants the Authority new functions and powers, including disciplinary ones, namely with regard to the whole anti-corruption strategy and transparency in the public sector. It also establishes the "concentration of functions": the Department of Public Service undertakes performance evaluation and measurement while the Authority performs the functions of corruption prevention and transparency. Law Decree no. 90 of 2014 is key and will reappear frequently in subsequent Authority documents.

11 July 2014
Appointment of Council members.
Four new members of the Council of the Authority are appointed by decree of the President of the Republic. They are Michele Corradino (an administrative judge) and Francesco Merloni, Ida Angela Nicotra and Nicoletta Parisi (university professors).

18 November 2014
The Authority's Resolution no. 146 of 2014 on "power of jurisdiction", which concerns the exercise of power of order by the Authority in the case of non-adoption of acts or measures required by the National Anti-corruption Plan or the Three-Year Plan to

Prevent Corruption, or by the rules on the transparency of administration, or in the case of conduct contrary to the plans and the rules on transparency.
The Authority can compel public administrations to carry out the commitments made in their Three-Year Plans to Prevent Corruption. However, its power of jurisdiction is not solely related to the Three-Year Plans. Nonetheless, the Authority needs to be careful here because it could run into criticism, and rightly so, because there are two key unwritten principles to bear in mind: (1) do not interfere with the work of the judiciary; (2) do not overstep the boundaries of the Authority's remit given its resources.

The Anti-corruption Authority has no mandate to—and indeed should not—interfere with the work of the judiciary. Having said that, in this regard the Authority makes for a useful observation platform. Via the Three-Year Plans to Prevent Corruption and through signalling (i.e. reporting), the Authority obtains much information of potential use to the judiciary. However, this is a by-product of the Authority's main processes; it has no involvement in judicial investigations.

The Anti-corruption Authority is keen to emphasize its distinctive "managerial sensitivity" to organizational management and organizational behaviour, a facet of its disposition not mandated by law. Although there are indeed anti-corruption provisions based on prohibiting unwanted behaviour (e.g. "revolving doors" or "pantouflage"), the heart of Authority's actions concerns objective organizational measures (penal law concerns the subjective responsibilities of individuals). Organizational measures are not the same as prohibition: they concern risk assessment, mapping administrative processes, and reorganization. In the National Anti-corruption Plan of 2016—and before that in the 2015 Update of the National Anti-corruption Plan of 2013—the Authority made it clear that the key is identifying organizational measures, implementing them and—it is to be hoped—checking their effectiveness.

31 January 2015
Deadline for the publication of the Three-Year Plan to Prevent Corruption 2015–2017.
Public administration's organizational units at this date still have time to publish their Three-Year Plan to Prevent Corruption 2015–2017, because their previous Three-Year Plan (2014–2016) is yet to be evaluated by the Authority.

1 February 2015
Evaluation of the Three-Year Plans to Prevent Corruption begins.
It will end on 28 October 2015 with publication of the 2015 Update of the National Anti-corruption Plan of 2013.

2 July 2015
The Authority's 2014 Annual Report to parliament. President Raffaele Cantone presents to parliament the first Annual Report from the National Anti-corruption Authority for the year 2014.

The 2014 Annual Report was written in 2015 and reports on the year 2014 and the first half of 2015; this is common practice in corporate annual reporting of comments on financial statements, and the same applies to independent authorities.

It is difficult to synthesize the final impact (or outcome) of a wide array of processes and ambitious goals such as those of the Authority. An effective entry point is the Authority's reporting process itself. As the reporting process was beginning in early April, one of the first questions was exactly how to report on the Authority's effectiveness.

As with all independent authorities in Italy, the Anti-corruption Authority is supposed to report to parliament rather than to the executive branch. Its report differs from the Anti-corruption Plan in that it concerns the Authority itself and its own effectiveness, whereas the Plan appertains to the anti-corruption efforts of public administration's individual organizations. The 2014 Report was the first by the "new" Anti-corruption Authority, previous ones having been presented by the its predecessors, CIVIT and AVCP (Fiorino, Galli and Petrarca, 2013). It is a 329-page document consisting of three parts:

1   The legal and institutional context of the new authority;
2   Public-sector procurement; and
3   Prevention of corruption and transparency.

The first part consists of three chapters. The first includes the Authority's proposed plan for its own reorganization following Law Decree no. 90 of 2014. The second includes the protocols of understanding for the prevention of illegality and the promotion of a culture of legality. The third deals with the Authority's international relations (with organizations at a universal level and at the European level, as well as bilateral relations with other states).

The second part of the report consists of six chapters, with subjects including the Authority's supervisory activities, extraordinary measures for the management of public contracts, and the activities of consulting, resolving disputes and market regulation.

The third part consists of four chapters and covers topics such as the areas of Authority intervention to prevent corruption and to promote transparency, the limits of current legislation, and proposals for amendments of such legislation. This part is relevant to the cycle of planning and feedback, mandated by the National Anti-corruption Plan.

14 July 2015
The first national meeting with the Corruption Prevention Officers.
This has the aim of initiating a meaningful dialogue on the issue of corruption prevention with specific reference to issues affecting those who hold the position of officer. Via a number of working groups, various themes are explored:

- Officers' relationships with political organs within their own organizations, bearing in mind the guarantees of independence and autonomy required in drawing up the Three-Year Plans to Prevent Corruption, the selection of measures, monitoring and control;
- Officers' powers and responsibilities, including cases of managerial responsibility and the necessary recognition of powers of coordination of the prevention strategy, especially in the relationship with the administrative structure;
- Relationships with the Authority (the protection of the officers by the Authority, collaboration with the supervisory activities carried out by the Authority, and future prospects).

One element that emerged from the meeting was the opportunity to link the anti-corruption plans to the performance plans under the aegis of CIVIT and the Department of Public Service, i.e. anti-corruption activities and provisions might become an element of substantive performance within the organization.

28 October 2015
The 2015 Update of the 2013–2015 PNA.
Every Three-Year Plan to Prevent Corruption is approved and remains valid for its three-year span. However, the plan can be updated annually. In 2015 an update was approved to the 2013–2015 plan. The new procedure that was initiated under Law Decree 90 of 2014 was that the update of the Three-Year Plan was to be written and approved autonomously by the Authority, without the Department of Public Service first making a proposal, as had been the case in the first release of the 2013–2015 plan.

The 2015 Update of the 2013–2015 National Anti-corruption Plan (PNA) (as approved and published in 2013) is a 53-page document, which includes an evaluation of the Three-Year Plans to Prevent Corruption published by the administrations before 31 January 2015, and also modifications of the Plan with a view to improving corruption-prevention measures.

The Three-Year Plans evaluated came from a total of 1,896 administrations (13 ministries, 67 national non-financial public agencies, 2 tax agencies, 67 state universities, 22 regions and autonomous provinces, 145 local health units, 90 hospitals, 20 scientific public institutions for hospitalization and treatment, 105 chambers of commerce, 110 provinces and 1,255 municipalities).

The main results of the evaluation are as follows:

- As of 28 February 2015, 96.3% of administrations had instigated and published at least one Three-Year Plan on their websites, and 62.9% had instigated and published the update for the 2015–2017 triennium.
- The quality of the plans was "widely insufficient".

- The quality of the plans is influenced by contextual variables such as the type of administration, the geographical location of the administrations and the organization's size.
- The quality of the plans shows significant improvement in the 2015–2017 programming relative to the 2014–2016 plans, indicating the presence of a learning curve owing to the gradual implementation of the legislation.

The 2015 Update of the National Plan ends with an explanation of the priorities and main objectives to be included in the Three-Year Plans to Prevent Corruption for 2016–2018, which administrations are to instigate and publish by 31 January 2016. These priorities and objectives are as follows:

- Transparency in the process of formulating the Three-Year Plan. Administrations must explain in the plan the actual process of its adoption.
- Connection between cognitive analysis and identification of measures. The measures taken must be based clearly and intelligibly on the results of the analyses.
- Centrality of the preventative measures. The authorities must clearly identify the measures to be taken and make their implementation explicit in the plan.
- Measures and responsibilities of the offices. Administrations must clearly spell out the concrete activities that the different offices will undertake in implementing the measures provided in the Three-Year Plan. It is essential that in the plan this division of labour comes within the scope of managerial responsibility or other performance measurement tools.
- Monitoring the effective implementation of the measures.
- Evaluation of the effectiveness of the measures taken.
- Integration between the Three-Year Plans and the transparency programme. The Three-Year Plan must contain the programme for transparency, which should show all the communication and publication requirements falling on each office, with the relevant responsibilities of the managers.
- Preventative measures and employees' duties of conduct. The adoption of a code of conduct is already required by Public Law no. 190 of 2012 and the PNA, but with regard to the evaluation of the Three-Year Plan contained in the 2015 Update, it is a requirement that the administrations undertake revisions of the codes adopted thus far.

The analysis of the Three-Year Plans has revealed that they are mostly treated as a mere formality rather than as a valuable tool for the prevention of corruption. The Update of the National Plan was prepared in any case, working on a combination of "simplification and differentiation of content" for the various administrations with the focus targeted on two risk areas only, such as, for example, health care and public contracts.

A key element in the preparation of the 2015 Update and the subsequent 2016 National Anti-corruption Plan was "the collaborative approach" which, by means of in-depth dialogues with external stakeholders, proved vital in acquiring sector-specific information and useful tips. The Update therefore introduced a sector-specific approach, including in-depth studies and suggested measures in the fields of public procurement and health care. The Authority planned to extend such case studies to other sectors in its future Three-Year Plans and Updates.

31 December 2015

The first Corruption Prevention Officers' report on the implementation of the Three-Year Plans to Prevent Corruption.

The officers' report is included in the evaluation and formulation of the National Anti-corruption Plan for the subsequent three-year period. The report must be made by 15 December of each year and concerns the implementation of previously published Three-Year Plans, up to 31 January of that same year. In the first evaluation round, in 2015, 190 plans were read; in the second, in 2016, 198 (PNA, 2016, 3 August, p. 9). It should be noted here that the new National Anti-corruption Plan in 2016 is also affected by Legislative Decree no. 97 of 2016.

31 January 2016

Deadline for publication of the Three-Year Plan to Prevent Corruption.

By this date the individual organizational units are required to publish their Three-Year Plans on their websites (PTPCT 2015–2017).

1 February 2016 (A)

Evaluation begins of the Three-Year Plans to Prevent Corruption.

1 February 2016 (B)

On this date a new plan for reorganization of the Authority was adopted, in response to new challenges it was facing. These challenges were confirmed by the new Public Procurement and Concession Contract Code—approved by Legislative Decree no. 50 of 2016—which identified the Authority as the supervisory and control system as well as market regulator, in addition to its responsibility for implementing measures to prevent corruption in public administration. The allocation of such new competencies required a response with regard to staffing and information systems, which inevitably implied higher costs.

However, these increased costs were somewhat at odds with the history of the Authority's budget. In fact, Law Decree no. 90 of 2014, merging the two authorities (CIVIT/ANAC and AVCP), requested the presentation of a reorganization plan by 30 December 2014, which was then to be subject to cabinet approval. However, it was not approved until February 2016; a critic might argue that two months would be sufficient for the approval process, rather than over a year.

Law Decree no. 90 of 2014, among its reasons for implementing a first reorganization plan, looked to reduce the Authority's operating costs by 20% (i.e. if 2014's costs were 100, those of 2015 should be 80 and remain so in future). This overt piece of housekeeping is not consistent with the Authority's new functions (including those added later) and limits its efficiency; for example, it blocks staff turnover—those who leave the agency cannot be replaced.

19 April 2016
The adoption of the new Public Procurement and Concession Contract Code—Legislative Decree no. 50 of 2016.
This includes at least a couple of noteworthy innovations. Tenders over €1 million no longer need to be assigned on a least-cost basis but will be awarded to the most efficient economic offer, which is to say the best price–quality ratio. This implies that the committee for the adjudication of the public tender can no longer include personnel from the procurement units, but members must be chosen from outside the organization and furthermore at random. A national approved list of potential members of such committees now has to be drawn up by the Authority, which is no small endeavour.

The Anti-corruption Authority will also define the criteria for selection of the procurement unit, which implies that not all current tendering units (about 37,000) will be allowed to put forth tenders. The Code envisages a reduction in the number of tendering units from 37,000 to less than 1,000 in the public works sector, while in the purchasing sector the tendering units have already been reduced to 35 central purchasing bodies. Of the 37,000 units, many do not put forth tenders for work, the most sensitive area, but just do procurement. The number of tendering units allowed to put forth tenders for work should also be reduced to 35.

The second noteworthy implication of this new legislation is the prohibition of the integrated tender. Let us explain this technicality. A work project goes through three different stages of design: preliminary, definitive and executive. Often, the executive stage is included in the tender, as procurement units do not have the capability to perform such a design phase. However, this integration of executive design and work led to a moral hazard, as executive design has a significant impact on the work's cost and characteristics.

24 May 2016
The second national meeting with the Corruption Prevention Officers.
The second national meeting with the Officers has the aim of discussing the many difficulties related primarily to the relationships with all those called to implement the measures provided by the National Anti-corruption Plan.

The figure of the Corruption Prevention Officer is presented not as a "policeman" tasked with enforcing the law, but as a key partner helping to embed the principles of legality and good performance of public administration as dictated by the Authority. Thus the relationship between the Authority and the Officers should be one of cooperation aimed at effecting a change in cultural perspective among public administrations.

The second national meeting collects the comments of the Corruption Prevention Officers with regard to the improvement of the new National Anti-corruption Plan. The new prevention strategy is not to be a requirement imposed by a distant and bureaucratic authority; rather, the Officers play a highly active role in fostering collaboration with the Authority in pursuit of the prevention of corruption and in the promotion of the principle of transparency.

25 May 2016

Review and simplification of the provisions in the field of prevention of corruption, publicity and transparency.

Legislative Decree no. 97 of 2016 amends the formulation of the National Anti-corruption Plan, after Public Law no. 190 of 2012. It also amends the Three-Year Plans which, with the addition of transparency provisions, become Three-Year Plans to Prevent Corruption and Promote Transparency, including as they do the previous Three-Year Plan on Transparency. The National Anti-corruption Plan 2016–2018 (approved in 2016) was written in accordance with this legislative decree.

Vis-à-vis previous legislation (Legislative Decree no. 33 of 2013), this delegated decree introduces two basic points:

1   A new transparency tool: civic access, applicable to all citizens, and to all data and documents held by the administrations. This proposal, which is typical of the Italian approach to freedom of information, is the subject of detailed regulation contained in the new Articles 5 and 5-b of Legislative Decree no. 33 of 2013. The new right of civic access refers fully to the freedom of information models: the right is granted to anyone and "is not subject to any subjective entitlement of the applicant"; the application for access need not be motivated, and the release of data and information is usually free of charge. In addition, the access tool—this time civic—is once again complementary to disclosure requirements: Article 8, c. 3 and Article 14, c. 2 clearly confirm this, in their reference to the use of the new instrument for data and documents the deadline for obligatory disclosure of which has expired (Carloni and Giglioni, 2017).

2   On anti-corruption legislation, this decree clarifies the role of the National Anti-corruption Plan in coordinating the preventative action of public administration and the stricter relationship between the Three-Year Plans to Prevent Corruption and the individual organizations' evaluation of performance.

22 June 2016
The first Italian Whistleblower Monitoring Report.
The first Italian monitoring on whistleblowing—"Reporting abuse and protection of the civil servant: Italy invests in whistleblowing, an important tool for the prevention of corruption"—was presented to the press at the Authority's headquarters. Over three years after the adoption of the rule protecting public servants who report abuses, the Authority has implemented a monitoring procedure on the state of whistleblowing in Italy in order to gather information on its occurrences and to understand to what extent the protection is actually perceived as a corruption-prevention measure.

The monitoring was undertaken on reports received by the Authority up to 31 May 2016 in order to identify characteristics about the reporter, the type of illegal conduct being reported and outcomes resulting from the reports. Some of the information obtained is presented in Exhibits 2.1–2.4.

In its first years of implementation the legislation has had to contend with the many regulatory deficiencies, and various aspects of critical issues that have led to the marginal use and/or malfunction of the whistleblower safeguarding tool. These include interference by the executives, prevailing political currents and widespread cultural resistance against the concept of

Exhibit 2.1 Number of whistleblower reports

	No. of reports	Average no. of reports per month
2014 (from 1 September)	16	4
2015	200	~17
2016	252	~21
2017 (until 31 May)	263	50

Source: ANAC, *Il whistleblowing in Italia, a cura di Anna Corrado*, 22 June 2017.

Exhibit 2.2 Reporting persons (whistleblowers)

Reporting persons	2014 (from 1 September) (%)	2015 (%)	2016 (until 31 May) (%)
Public servant	69	60	71
Private	6	1	9
Corruption Prevention Officer	6	9	6
Anonymous	13	6	3
Executive	–	12	7
Municipal police	6	3	2
Municipal and provincial councillors	–	6	2
Military	–	3	–

Source: ANAC, *Il whistleblowing in Italia, a cura di Anna Corrado*, 22 June 2017.

Exhibit 2.3 Relevance of whistleblower reports

| | Relevance of reports | | |
	Low (%)	Medium (%)	High (%)
2014 (from 1 September)	81	19	–
2015	47	30	23
2016 (until 31 May)	42	33	25

Source: ANAC, *Il whistleblowing in Italia, a cura di Anna Corrado*, 22 June 2017.

Exhibit 2.4 Kinds of whistleblower report

Reports	2014 (from 1 September) (%)	2015 (%)	2016 (until 31 May) (%)
Failure to implement anti-corruption laws	–	2.7	1.1
Environmental damage	–	0.8	–
Poor management of public resources	–	7.0	2.2
Abuse of power	–	7.0	5.6
Conflict of interest	5	5.1	7.8
Illegitimate competition	–	6.6	6.7
Illegitimate procurements	5	18.3	21.1
Illegitimate appointments and nominations	10	10.1	12.2
Loss of revenue	5	1.2	–
Misappropriation	5	–	–
Corruption and maladministration	50	23.3	27.8
Loss of position and illegitimate transfer arising from reporting	20	–	15.6

Source: ANAC, *Il whistleblowing in Italia, a cura di Anna Corrado*, 22 June 2017.

whistleblowing, so manifest that it remains difficult to find an Italian term that does not have the negative connotation of "informer" or "spy". Thus far, reporting, as far as public administrations are concerned, rather than being a risk minimization tool, is still perceived as an actual risk factor and there remains a faction of public servants who view illegal conduct and poor management not as a deviation but as an innate part of administrative behaviour. In this context, encouragement of whistleblowing is complicated. To overcome such difficulties, it is vital that the top levels in each administration invest heavily in the implementation of the legislation at least in the medium term, superseding the emergency-management nature of whistleblowing and establishing structural and effective management in its stead.

While administrations struggle to make their internal environments more conducive for whistleblowing, the Authority continues to provide support. It fills the regulatory gaps by issuing guidelines, providing guidance on interpretation and finding operational solutions that simplify the procedure for managing the whistleblowing process. Among the many significant initiatives taken up to 2017 are:

- The strengthening of the "Whistleblowing Working Group" with the Financial Crimes Police;
- The development of a specialized IT platform, which is licensed gratis to the administrations and which they can adapt to their specific needs;
- The stipulation of protocols of agreement with Libera Association and Transparency International–Italy and with the Attorney General's Office at the Court of Cassation;
- The initiation of dialogue channels, such as the National Day of the Compliance Officer, held at the Bank of Italy's conference facilities;
- Protocols with public administrations to manage earthquake relief processes.

14 July 2016
The Authority's 2015 Annual Report to parliament.
This is an activity report, detailing the many strands of the Authority's action. It is a 360-page document comprising three parts:

1   The Authority and its context;
2   Prevention of corruption and transparency; and
3   The public procurement sector.

The first part consists of two chapters and describes the Authority's new powers (following new legislation: Legislative Decree no. 50 of 2016; Legislative Decree no. 97 of 2016; and Public Law no. 69 of 2015), the Authority's Reorganization Plan, the corruption-prevention tools and the Authority's international relationships (with organizations at a universal level and at the European level, as well as bilateral relations with other states) (Parisi, 2017).

The second part consists of three chapters, with subjects including the path towards the 2015 Update of the National Anti-corruption Plan 2016–2018 (approved 28 October 2015) and actions in the field of corruption prevention (regulating, supervising and sanctioning activities) and regarding protection of transparency (regulating and supervising activities).

The third part consists of eight chapters and covers topics such as the market for public contracts, the new "collaborative" supervision, supervision of the qualification system and sanctioning activity, controls and the extraordinary measures related to public procurement, regulatory activities, advisory activities, and arbitration and the activities of the Arbitration

Chamber. The collaborative supervision should be defined as a particular and exceptional form of verification, above all preventative, aimed at fostering productive collaboration with the contracting authorities and thus guaranteeing the correct functioning of tendering processes and contract execution, while also impeding criminal infiltration attempts in the tenders.

This is the second report to parliament presented by the "new" Authority. It comprises an evaluation of the complex process of building an effective system of corruption prevention in Italy. In its introduction it highlights a few key measures of activity and we will cover three paragraphs from page 1 of the report in order to offer an account of how the Anti-corruption Authority measures its own output, i.e. which activities and which "products". The report to parliament focuses on the supervision activity of the Authority, which has an increasing impact on the structure and organization on the Authority itself. Exhibit 2.5 summarizes the data provided on page 1.

Such a table can be thought of as a "dashboard" for the Authority: a few numbers the President keeps handy in his pocket. An example of this kind of reporting can be found in La Noce (2015). Let us explain in detail exactly what those activities and numbers mean. The first measurements to consider concern reports received from the field. As seen in Exhibit 2.5, the total of complaints received is 5,895.

The Authority received 2,990 complaints combined on matters of public works (1,660), and service and supplies (1,330). This figure represents an increase of over 50% compared to 2014 (when 1,200 complaints were received). Moreover, over 1,500 reports were received regarding qualification of companies and a further 1,435 came from the Transparency Campaign web platform, submitted in the period 1 January–31 December 2015 (+90% compared to 2014). This web platform is designed to receive complaints about the transparency of the institutional websites of public administration organizations. These 1,435 reporting forms concerned 542 public-sector organizations (+59% compared to 2014). The forms were presented either by individual citizens (81% of cases, +13% compared to 2014), by associations (13%, –3% compared to 2014) or, in only 4% of cases (–10% compared to 2014), by public administration organizations themselves.

Analysis of the contents of the forms revealed that 16.5% of cases detected a total absence (7%) or general deficiency (9.5%) of the "transparent administration" requirements of institutional websites, a slight increase from the 15.5% recorded in 2014. The remaining 83.5% of the reports were about alleged violations of specific duties of publication (i.e. specific information missing from the websites that was supposed to be there).

Sector analysis of the forms reveals that the largest number of reports concerned "tenders and contracts" (9%), double compared to 2014, followed by obligations related to "political bodies" (8%). There has been a significant increase from the previous year of reports related to violations concerning the failure to publish the procedures for the exercise of the right of civic access,

Exhibit 2.5 Possible "dashboard" for the Authority

Items/activities	2014	2015	Δ%	Chapter	Page
Complaints received/anomalies					
Public works		1,660		7	178–9
Service and supply		1,330		7	178–9
Sub-total new supervision	1,200	2,990	>50	7	178–9
Qualification of companies		1,500		Introduction	1
Transparency campaign		1,435		5	135
Total input/reports from the field		**5,895**		–	–
Cases/investigation started					
Public works, service and supply		1,880		7	178–9
Special supervision		600		7	178–9
Company qualification system		2,560		9	230
Corruption prevention		929		4	97
Transparency rules		341		5	127
Sub-total corruption and transparency		1,270		–	–
Total output/cases started/finished		**6,310**		–	–
Counsel					
Art. 32 dl. 90/2014		47		Introduction	1
Inspections		41		Introduction	1
Counsel/opinions		>940		Introduction	1
Total counsel		**>1,028**		–	–

Source: elaboration on ANAC, Annual Report to Parliament for 2015.

of information on the prevention of corruption measures, data on public works, databases and environmental information.

In general, an analysis of the demand for Authority intervention showed a shift from personal data of a financial or personal CV nature to data related to administrative activity associated with prevention of corruption. As regards investigations following these complaints, we can affirm that the exercise of the Authority's supervision was strengthened through inspections and special supervision activities. The inspections are intended to identify anomalies which will then be the subject of investigation. The special supervision activities are designed to carry out verifications on a wider spectrum of sectors or cases considered particularly relevant (e.g. the procurement activities of the Milan Expo 2015). The number of cases processed by the Authority totalled 6,310.

The Anti-corruption Authority activated 1,880 files following the 2,960 complaints received about public procurement, services and supplies, while special supervision activities led to over 600 investigations. Such numerous investigations are being carried out by the regular law enforcement bodies, such as the Financial Crimes Police.

Much activity was devoted to monitoring the actions of certification bodies: those that verify that companies who want to enter a public tendering process and perform public works have the required qualifications. The overall proceedings undertaken in the context of supervisory activities over the qualification system came to about 2,560 units, and in 154 cases led to the imposition of fines of €290,830 in total.

The Authority also initiated 1,270 proceedings: on the application of measures to prevent corruption (929) and legislation on transparency (341). The objective of this activity was to provide feedback on the Three-Year Plans written by the thousands of organizational units of Italian public administration.

Vigilance following reporting was the Authority's single most time-consuming activity of 2015. On the matter of prevention of corruption, 929 proceedings were initiated, of which:

- 115 dealt with the discrepancy between the preventative measures taken by the supervised institutions and actual administrative behaviour, and gave rise to the initiation of proceedings for the adoption of measures aimed at ensuring compliance by the authorities of the measures to prevent corruption;
- 54 related to proceedings concerning the existence of any causes of disqualification, ineligibility and unfitness for office (Carloni, 2017);
- 54 were sanctioning proceedings initiated in relation to the regulations on the exercise of the sanctioning power of the Authority.

Approximately 42% of reports concerned municipalities; the rest, for the most part, concerned other kinds of local authorities. The reports mainly related to professional assignments, career advancement, conflicts of interest, staff rotation or the failure to adopt Three-Year Plans or Codes of Conduct. The regions generating the highest number of reports were: Campania (19.5%), Lazio (12.9%), Sicily (10.8%) and Puglia (7.9%).

Activities in the field of transparency led to a total of 341 proceedings, of which 174 were regulatory proceedings and 167 sanctioning proceedings (see Exhibit 2.6). Of the regulatory proceedings, 135 were initiated following reports and 39 were initiated ex officio, as reported. With regard to sanctioning proceedings, 110 requests for information relating to violations subject to administrative fines were activated.

To complete this description of the measures of the Authority's 2015 Annual Report, it is necessary to take note of the figures related to extraordinary measures pursuant to Article 32 of Law Decree no. 90 of 2014 (47), inspections (41), and opinions given in the context of counselling functions (over 940).

A first conclusion of our description of the Authority's 2015 Annual Report is that we notice that only a few measures reported are relevant to the cycle of planning and feedback. We therefore obtain a view that the annual cycle of planning and feedback does not bring results per se but is a catalyst for specific action, taken on the basis of the many further powers of the Authority. The cyclical process therefore is a basic theme. The Anti-corruption Authority would go on from here implementing its iterative adaptation process and the expected next step was the new National Anti-corruption Plan (2016–2018, approved 2016).

3 August 2016
The new National Anti-corruption Plan PNA 2016–2018.
This new plan, a 100-page document, impacts organizations' Three-Year Plans to Prevent Corruption 2017–2019. The first 30 pages are on general

Exhibit 2.6 Proceedings concerning transparency

Type of proceeding	No.
Regulatory proceedings:	
initiated ex officio	39
initiated following reporting	135
Total regulatory proceedings	174
Sanctioning proceedings:	
requests for information for possible initiation of sanctioning proceedings	110
notices of initiation of sanctioning proceedings against political office-holders	57
Total sanctioning proceedings	167
Total proceedings concerning transparency	**341**

Source: ANAC, Annual Report to Parliament for 2015, p. 127.

matters but a specific element is the downplaying of rotation as one of the anti-corruption measures public administration organizations are invited to take. The subsequent 70 pages, consolidating the trend inaugurated by the 2015 Update to the Plan, include specific sectoral guidelines. The sectors here are: land management, health care, protection of cultural heritage for the public good, small municipalities (there are 8,100 municipalities in a country of 60 million inhabitants), professional charters and metropolitan cities. In view of the poor quality of the Three-Year Plans to Prevent Corruption, the Authority decided against "one size fits all" guidelines and chose to be specific in a bid to improve the quality of the Three-Year Plans to Prevent Corruption and Promote Transparency.

We hope the above offers a sufficient understanding of the Authority's ideas and operations, allowing us to reflect on and learn from the Authority's experience. We will refrain from going into any further detail to avoid over-complication; furthermore, we would prefer to offer an easily digestible summary.

The story of the Anti-corruption Authority does not, of course, end here. The 2017 Update to the 2016–2018 National Anti-corruption Plan included specific guidelines about the implementation of EU policies, waste management and universities. Legislation or regulation is also in the offing to reduce the number of state or municipally owned companies from the current total of 7,000 to no more than 1,000.

2.4 Conclusions

This chapter has presented an exposé of the Italian Anti-corruption Authority (ANAC)'s cycle of anti-corruption measures and its duties relating to corruption prevention. But it should be made absolutely clear that there are many other forms of intervention of a preventative nature which require the efforts of different actors: a more efficient and less invasive bureaucracy; honest, authoritative and credible politics; a business world that, as occurred in the struggle against the mafia, chooses to be on the right side—all of these would make an invaluable contribution to the prevention of corruption.

The Authority's fundamental task remains a highly challenging one. The contexts in which it operates in the struggle to limit corruption are strategic, but the tools at its disposal require time and institutional collaboration to take effect, because nobody should be under the impression that they are the bringer of miraculous panaceas.

This cycle of planning and feedback is in itself an acknowledgement that corruption must be tackled through organizational and incremental change. The process of implementing anti-corruption measures must rely on a mix of voluntary organizational change and the law. Repressive penal law cannot address every minute detail. We infer the following concepts: organization is different from repressive penal law; organization is at least to some degree more detailed than penal law; corruption depends on organization and organizational choice.

The Authority works under different terms from those of the classic mechanism of penal law. Penal law prohibits certain individual, subjective behaviour, e.g. "thou shalt not steal". Penal law is an order given to individual public civil servants. The Authority uses a different logic: it works via objective measures, tackling corruption without taking into account the subjective propensities of individual civil servants. This is a more difficult and complex approach, but one which can deliver results in the long run. The Authority tries to convince public administrations that anti-corruption measures improve performance as regards their actual missions. In this sense, the Authority focuses on the organization of public administrations in order to encourage implementation of the law.

Our short analysis of the Authority's 2015 Annual Report to parliament offers a glimpse of the many further activities it performs that are adjacent to, and sometimes dominant over, the corruption prevention cycle. The next chapter gives a description of the Authority's organizational chart and takes an extended look at its activities.

Chapter 3

An inside-out view of Italy's Anti-corruption Authority

Thus far we have described the activities of the Italian Anti-corruption Authority (ANAC) in the form of a timeline, acknowledging the statutes that have over time come to establish and regulate the Authority. In this chapter, to gain a wider perspective on the Authority, we will review its activities in a systematic way using its organizational chart.

ANAC is a collegial independent authority, composed of five members appointed by the decree of the President of the Republic, acting on a proposal by the Prime Minister and the Minister for Public Administration. The Authority is now also responsible for public contracts, a responsibility previously conferred on separate supervisory authorities.

The top organizational unit and decision-making body of the Authority is the Council. The Council is in fact often taken to be the Authority. It comprises five positions: President and four other members. As shown in Chapter 2, the members of the Council were appointed in 2014 and will remain in office until 2020. The President is Raffaele Cantone, a public prosecutor who, at the time of his appointment, was working at the Court of Cassation having done major work against the mafia.

The individual members of the Council are: Michele Corradino, Francesco Merloni, Ida Angela Nicotra and Nicoletta Parisi. Michele Corradino was a judge of the State Counsel; Francesco Merloni was a professor at the University of Perugia; Ida Angela Nicotra was a professor at the University of Catania and a member of the Joint Commission State-Region of Sicily; and Nicoletta Parisi was a professor of International Law at the University of Catania. She was also the appointed expert of the Italian government in the Public Prosecutor's Office at the European Commission.

The Authority's organizational chart is basic (see Exhibit 3.1): the Council and the President are the governing body. Auxiliary bodies include the overseen accountants. The Secretary General is the head of the operating structure. The structure is divided according to supervision and regulation activities. The President, the Council and the Secretary General also have their own staff.

The Authority's structure is a result of the merger of two previously existing bodies: the Commission on Evaluation, Integrity and Transparency

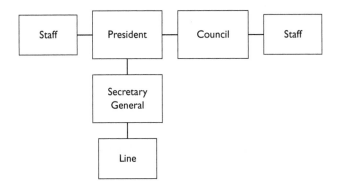

Exhibit 3.1 The Authority's organizational chart.
Source: elaboration on Italian Anti-corruption Authority (ANAC) documents.

(CIVIT) and the Authority for the Supervision of Public Contracts (AVCP). CIVIT had a nucleus of 30 employees, 15 of whom were functionaries (middle-level employees). AVCP had 330 employees of whom 50 were managers. Of these, 25 were demoted to the rank of functionary after the CIVIT–AVCP merger.

At the end of 2014, the Authority produced its own reorganization plan (*Piano di riordino*) which included a reduction in the number of managers, as described above. According to the plan, the Authority's budget was to be limited to €52 million annually.

Returning to the Authority's organizational structure, a complete list of staff reporting to the President is as follows:

- Secretary and staff of the President;
- Secretary and staff of the Council;
- Technical secretary;
- Special operations units;
- Litigation (bureau);
- Pre-litigation[1] and legal affairs;
- Guidelines, general deliberations and supervision indicators;
- Inspections;
- Supervision plans and special supervision.

Staff reporting to the Secretary General are:

- Protocol, document flow and decision-making support;
- Secretary and staff;
- Human and financial resources;
- Information systems operations;

- General services, tenders, contracts and logistics;
- Design and development, information systems and the Authority's web portal.

Line bureaux reporting to the Secretary General are:

- Functions of supervision;
- Functions of regulation.

We will now examine the latter two in greater detail, taking them in order.

3.1 Supervision activities

Broadly, the Authority is responsible for the preparation of the National Anti-corruption Plan, the definition of guidelines for codes of conduct, and the supervision of the adoption and the effective implementation of anti-corruption instruments, beginning with the monitoring of compliance with transparency obligations and the supervision of public tenders.

We will continue by describing the supervision activities.

3.1.1 Supervision on corruption prevention

In this section we describe activities that have already been touched on in Chapter 2 with a view to providing greater detail. The Authority supervision activity forms the very bedrock of the Authority's statute (Public Law no. 190 of 2012). This statute makes a distinction between corruption prevention and corruption repression, and dictates that every public administration organization must draw up an anti-corruption plan. Such a plan must include organizational provisions to prevent corruption or to contain the impact of corruption.

Public Law no. 190 of 2012 also mandated the establishment of the National Anti-corruption Plan. We may recall here that the National Anti-corruption Plan adopted a very broad definition of corruption, basically viewing corruption as "maladministration".[2] The individual organization's anti-corruption plan, then, is to be understood as an instrument to help contain corruption over the long run, i.e. corruption in its wider sense of maladministration.

The National Anti-corruption Plan is structured

> as a programmatic tool subject to an annual update with the inclusion of indicators and targets for corruption in public administration in order to make the strategic objectives measurable and to ensure the monitoring of possible divergences from these targets arising from the implementation of the PNA.

> (ANAC, *National Anti-corruption Plan*, 3 August 2016)

Guided by the national plan, each public administration identifies, via its own plan, the specific risks of corruption in its own administration and the measures deemed necessary to tackle them.

The PNA remains the heart of the various measures of the anti-corruption strategy and brings together various documents (the Three-Year Plans for Transparency and the Code of Conduct) and integrates them as part of a system along with other organizational measures, as stipulated by the law. These can also indirectly help raise standards of conduct, for example through an overall improvement in public performance (the performance plan) or digitization (the digitization plan).

The Three-Year Plan to Prevent Corruption constitutes the essential point of reference for each administration—on the one hand in the drafting of anti-corruption policies, and on the other in adapting them to the specific context and its effective level of risk. The approach taken is essentially as follows. Each administration has to assess the level of risk of corruption for each sector in which it operates: some sectors have already been identified as "high-risk" by the national plan and the legislation (staff recruitment, contracts and procurement, concessions and economic subsidies). It is the responsibility of each administration to conduct its own internal analysis and establish the most appropriate administrative mechanisms (transparency, staff turnover, procedural rules, employee obligations, digitalization of procedures, etc.) to address the identified risk. This is, therefore, a collection of preventative measures. In the case of an incidence of corruption, the Corruption Prevention Officer and the particular administration will have to demonstrate to the Authority that appropriate prevention measures had been put in place and that the corruption incident is therefore an extraordinary and unpredictable event (which, nonetheless, justifies a further tightening of the preventative measures).

The national plan, as defined by the Authority, is subject to annual updates, as is each administration's Three-Year Plan. However, the national plan has a fixed three-year span and can be updated in any year of those three, whereas the individual organizations' plans have a "moving" three-year span, which is to say that they are revised every year, continually looking forward three years, so there are no "updates" as such.

The national plan comprises a main text (56 pages) and five annexes (88 pages). The document defines the strategy for prevention of corruption, at national and decentralized levels. At the national level, it outlines a definition of corruption and establishes the Anti-corruption Authority's strategic objectives. Its concept of corruption is a broad one and includes all situations in which, in the course of administrative activities, "a public employee abuses the power entrusted to him or her in order to obtain private benefits" (Section 2.1). It also specifies that the notion of corruption is broader than specific instances of penal crimes, including the "malfunctioning" of public administration because of use for private ends of the entrusted functions. The definition also encompasses the "mere attempt to pollute administrative action from outside". Such a wide-ranging definition is what has led to the under-standing of corruption as "maladministration" (Carloni, 2018).

The formulation of the national anti-corruption strategy aims to achieve the following strategic objectives, with a clear "cascade model" of action:

- To reduce the opportunities for corruption to occur;
- To increase the capacity to detect corruption;
- To create an unfavourable environment for corruption.

The main tools proposed by Public Law no. 190 of 2012 in pursuit of tack-ling corruption are as follows (PNA, 2013, 24 July).

3.1.1.1 Adoption of Three-Year Plans to prevent corruption

The Three-Year Plans identify, on the basis of the National Anti-corruption Plan, the specific risks of corruption in each individual administration and the measures deemed necessary to address them. Each public administra-tion's Corruption Prevention Officer assists their organization in drafting and publishing a Three-Year Plan to Prevent Corruption. The Officer's role is to involve the whole organization, from top to bottom, political bodies included (which is to say those bodies with political responsibilities: the politicians who may have a top-executive role in the organization, as is the case with the executive ministries or the presidents of SOEs, who are appointed by politicians). The Three-Year Plan is drawn up according to the instructions given in the national plan and is presented to the politicians for approval. In the 2015 Update of the 2013 PNA, it was asserted that the anti-corruption effort would be ineffective without the politicians' engagement.

The Three-Year Plan must contain a corruption risk analysis and proposed organizational changes designed to prevent corruption. Such changes are sug-gested within the national plan and are described in the following paragraphs. Thus a combination of prescription and flexibility affords consistency of the system at a national level yet allows individual public administrations a degree of autonomy to apply the most effective solutions of their own choosing.

The duty of the Officers to involve and make politicians aware of their responsibilities is an important step, because in Italy the distinction between politics and administration (Wilson, 1887) was "juridicized" and made into Public Law no. 29 of 1993 (Merloni, 2006). Now the requirement for both politicians and bureaucrats to be engaged in anti-corruption work appears somewhat to contradict the need for a distinction between politics and administration. The Officer is a central figure in the corruption prevention system with significant responsibilities, as well as having the position of being a privileged interlocutor of the Authority, which confers a number of important functions. He or she has the crucial task of proposing the adoption of the Three-Year Plan to the political bodies governing their individual public administration organizations, of verifying its correct implementation and its continuing suitability, as well as reporting the results of the activity at the end of each year. His/her obligations also include reporting to the judicial authorities any ensuing corruption cases.

3.1.1.2 Obligations of transparency

The Three-Year Plan to Prevent Corruption will include the Three-Year Plan for Transparency and Integrity, established by Legislative Decree no. 150 of 2009.

3.1.1.3 Codes of conduct

Each public administration will adopt its own code of conduct, characterized by a concrete approach, readily understood by employees and taking into account the directions of the Anti-corruption Authority, namely:

- Include compliance with a code of conduct as a condition in the text of contracts and tenders with business firms (or economic operators [OEs]);
- Plan appropriate internal training initiatives on the code;
- Verify the suitability of the organization's procedure for personnel disciplinary proceedings.

3.1.1.4 Personnel rotation

Three-Year Plans to Prevent Corruption must include measures to realize the rotation of managers and employees operating in areas at higher risk of corruption.

3.1.1.5 Duty to abstain in cases of conflict of interest

Three-Year Plans to Prevent Corruption must include actions for public administrations to inform their personnel and top managers about the obligation to abstain from positions, appointments or jobs in cases of conflict of interest (moral hazard). Public administrations must also inform their personnel about the consequences of violation and about the correct procedure to follow in cases of conflict of interest.

3.1.1.6 Specific guidelines on the performance of official duties, activities and extra-institutional mandates

Complementary to the action above regarding conflict of interest, public administrations need to define *ex ante* criteria for the acceptance by their own personnel and top managers of appointments or jobs outside their own public organization.

3.1.1.7 Specific guidelines on the implementation of the new legislation on appointments to management positions

These guidelines focus on public managers' career paths and the measures apply to three different stages: the managers' positions before they are appointed; their positions while they are in office; and their positions once they leave office. This is specifically mandated by Legislative Decree no. 39 of 2013.

Public administrations need to ensure that all future management appointments take into account any past activities of individuals that might be in conflict of interest with the vacant position. As such, they need to specify criteria for incompatibility with management positions in order to obviate conflicts of interest.

This provision aims to prevent what in bureaucratic jargon is called the "revolving door", i.e. managers moving frequently and freely between political and management positions. Addressing the revolving-door phenomenon implies the notions of disqualification, ineligibility and unfitness for office, or *inconferibilità*—a new concept introduced by the law, the aim of which is to exclude from public office those who find themselves in a situation that puts at risk the integrity, or even the appearance of impartiality, that should characterize public conduct and civil servants (Carloni, 2017).

While they are in office, managers should abstain from decisions in which possible conflicts of interest may arise.

Specific guidelines apply to activities subsequent to the termination of employment. The issue addressed here is "pantouflage", which refers to corrupt appointments of managers whose positions in public administration have been terminated or from which they have retired. Public administrations are required to impose internal guidelines to prevent employees and managers from accepting positions with third parties that are in conflict of interest with their terminated position (Andracchio, 2016).

There are guidelines also on the formation of committees, assignments to offices, or appointments to management positions in cases of convictions for crimes against public administration. Those with convictions should not be considered for positions within public administrations.

3.1.1.8 Specific guidelines on the protection of employees reporting illegalities (whistleblowing)

Protection is afforded by the introduction of confidentiality provisions in the Three-Year Plans to Prevent Corruption, and each administration is to provide internally diversified and confidential channels for receiving reports, the management of which is to be entrusted to only a few individuals. In addition, data revealing complainants' identities must be encrypted.

Whistleblowing is a proposition that is struggling to gain traction because the prevailing protections offered to the whistleblower are not believed to be effective and because there is little inclination to report unlawful behaviour (reporting is often viewed as "informing" in the sense of betraying to the authorities). In order to stimulate more frequent use of this measure within public administrations, the Anti-corruption Authority, following a wide-reaching public consultation, published ad hoc guidelines—Resolution 6 of 2015—which provide the administrations with recommendations on how to adequately protect whistleblowers while creating awareness of the necessity of systems of protection.

More recently, parliament has approved a reform of whistleblowing legislation with Public Law no. 179 of 2017.

3.1.1.9 Planning on ethics, integrity and other issues related to the prevention of corruption

Public administrations should plan specific actions in this regard in their Three-Year Plans to Prevent Corruption.

It may be noted that six of the nine measures listed above focus on ensuring the legitimacy of those holding office in managerial positions. Provisions of this nature are viewed by the Authority as key elements of organization within public administration.

3.1.2 Cycle of planning and feedback

The system must ensure that national strategies are developed and modified according to the needs of the administrations, and in light of feedback received from them, as the targeted prevention tools are gradually put in place. According to this logic, the adoption of the National Anti-corruption Plan is not to be seen as a one-off but as a cyclical process in which strategies and tools are gradually refined, modified or replaced in light of feedback obtained about their application. In addition, the adoption of the plan takes into account the progressive development of the system of prevention itself, with an understanding that success depends largely on a consensus about preventative policies, on their general acceptance, and on all actors involved getting genuinely behind them. For these reasons, the national plan is aimed mainly at facilitating the full implementation of legal measures, i.e. those tools for prevention of corruption that are mandated by the law.

The national plan is a document addressed to all public administrations (in accordance with Article 1, Paragraph 2 of Legislative Decree no. 165 of 30 March 2001) and to the wider "public sector", including SOEs. The bounds of its application will be illustrated in more detail in Chapter 5, "The social and economic context of the anti-corruption effort". The national plan mandates that every public administration updates its own Three-Year Plan to Prevent Corruption annually by 31 January, with the ultimate aim of adapting the anti-corruption strategy to changing organizational conditions and the internal and external environment.

The basic concept of planning and feedback between the National Anti-corruption Plan and the Three-Year Plan to Prevent Corruption, and between public administration and the Authority, is illustrated in Exhibit 3.2 as an iterative and cyclical process. The Authority's mission is to make the process converge towards containment of corruption nationally.

It is clear that the system's feasibility lies in its flexibility in adapting to the specific requirements of each administration, and in its careful processes of analysis and adjustment. An alternative would be a more formal system in which plans emerge from isolated work in a limited number of offices and are then simply copied from others' experiences and documents, and are poor in terms of their analysis of context and risk assessment (ANAC, *Annual Report*, 2016).

In particular, complying with the bare minimum of requirements, i.e. merely identifying risk areas specified as compulsory by the national plan, carries with it the risk of ineffective treatment owing to a lack of comprehensive analysis and broad-spectrum diagnosis.

Furthermore, the idea of a public administration being exposed to the risk of corruption only when in "giving" mode (i.e. recruiting or promoting staff, assigning works or contracts, recognizing contributions or non-economic benefits, etc.) is a restrictive one, especially in relation to the nature of certain administrations. The administration is subject to risk even in penalizing an offender, particularly if formal or informal trading begins in response to the dispute. This is one reason why each administration is responsible for establishing a Three-Year Plan tailored to its own context.

For the sake of completeness, mention should be made here of the fact that anti-corruption measures are leading to a more considered approach to relationships with the representatives of organized interest groups, albeit in still limited terms: while lobbying is a phenomenon still awaiting comprehensive regulation in Italy, it is touched upon by the National Anti-corruption Plan and is also affected on several fronts by the anti-corruption legislation (e.g. rules governing public contracts and the illegal traffic of influence) (David, 2017).

The Anti-corruption Authority, in monitoring anti-corruption measures, is similar to the US Office of Government Ethics in that it does not deal in repressive penal measures. However, unlike the Office of Government Ethics, it does deal with complaint reports about misconduct and follows up with investigations, but only within its purview of checking the quality of the adopted measures.

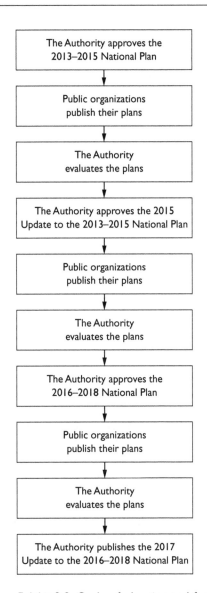

Exhibit 3.2 Cycle of planning and feedback.

The Authority checks what public administrations are actually doing to prevent corruption: it assesses the Three-Year Plans, including in terms of their "quality" (by means of spot checks). Its annual reports describe a number of persistent problems in the anti-corruption policies of individual public administrations, often caused initially by a poor assessment of the internal and external context, with plans often being drafted by a few officials working in isolation from the political leadership (the latter being responsible, nonetheless, for the plan's formal adoption) and with ensuing documents that aim for formal legal compliance, operating both of out of the organization's own initiative and on complaint reports from the field.

One way of understanding the relationship between the Authority's initiative and reporting is to divide activities into: (A) cyclical activities and (B) specific cases. Standard activities in the (A) category include:

A.1 Conflicts of interest;
A.2 Official professional associations; and
A.3 Appointments.

The (B) category (specific cases) includes:

B.1 The anti-corruption plan of the City of Rome;
B.2 The annulment of appointments; and
B.3 Whistleblower reports.

We will now describe these in more detail.

A.1 Conflicts of interest

The Authority has intervened several times in conflict-of-interest situations. This occurs when there is an appointment to a decision-making position in an area in which the appointee has personal or professional interest, a situation that runs counter to the impartiality that such an appointment requires.

To address this, Presidential Decree no. 62 of 2013 has been enacted, which contains a list of duties relating primarily to the extent of the (potential or actual) conflict of interest, and which result in reporting and transparency requirements, obligatory abstention (from participation in decision-making) and the disclosure of the interests involved. This is applicable to all employees, specifically targets conflicts of interest, and tackles corruption by favouring disclosure (with transparency rules and reporting obligations), procedural rules (legislative action regulating the procedure) and obligations and prohibitions. As such, this code of conduct is an important part of an overall strategy to reduce the risk of maladministration.

A.2 Official professional associations

It first of all needs to be noted that the Authority's resolution pertaining to official professional associations—being public bodies and therefore within its orbit of supervision—was challenged in court on the grounds of whether it legitimately had jurisdiction in this area. The court ruled in favour of the Authority (in September 2015), which was henceforth able to begin its supervision of the governing bodies of official professional associations in 18 Italian cities. On examination of their websites it found that none complied with rules concerning the adoption of the Three-Year Plan to Prevent Corruption, the Three-Year Transparency Plan or the Code of Conduct. All bodies were asked to update their websites in accordance with the law, and a follow-up revealed that almost all had complied. Some cases required changes in the associations' regulations which failed to include publication of income and financial information. The Authority held meetings and gave assistance to the associations in complying with the transparency regulations. In some cases, the websites subsequently adopted open data standards, having formerly allowed access only to their own members.

A.3 Appointment of personnel from outside the organization

The Authority also acted on complaints of misconduct reported by Corruption Prevention Officers, civil servants and citizens concerning the hiring of persons from outside public administration. The targets of these investigations were public organizations of local communities, often municipalities. The law has been updated and now, quite rightly, specifies that local authorities must abide by certain rules in hiring persons from outside public administration for consulting or other kinds of collaborative work.

Let us now turn to specific cases (our category B).

B.1 The City of Rome's anti-corruption plan

In 2015, the Authority undertook a detailed investigation into the activities of the Municipality of Rome, Italy's capital, scrutinizing its 2015 Anti-corruption Plan. Attention focused on the organizational units of procurement and personnel, the judiciary having intervened in these areas. An inspection report was drawn up which revealed a number of irregularities in the Municipality's activities. The Authority asked the Municipality to produce an updated report on the measures it would take in the course of 2015 to address those irregularities—specifically, what changes were to be made to its statute and organizational structures. The city was also asked to make changes to its internal regulations on procurement, and to simplify and increase transparency in its administrative processes. Finally, the Authority's report asked the Municipality of Rome to produce an adequate reporting tool to monitor its procurement activities.

B.2 Annulment of appointments

The Authority revealed that the 2015–2017 Anti-corruption Plan of the Municipality of Rome was lacking in corruption-prevention measures in the area of recruitment. There were no objective mechanisms to verify the professional competence of candidates with regard to the vacant positions. The plan had no specific anti-corruption measures for the organizational processes of "personnel recruitment" and "promotions and salary increases" beyond the measures suggested by the National Anti-corruption Plan. Finally, no follow-up was catered for. The Authority requested that the Municipality of Rome abide by the Authority's report in drawing up its future anti-corruption plans.

B.3 Whistleblower reports

The Authority is also responsible for managing whistleblower reports (*segnalazioni*) from public administration employees. Such reports may involve both anti-corruption activity as well as supervision on public procurement, which fall to different units of the supervision organizational area. In response, the Authority established a permanent working group which draws members from different organizational units and also includes a member of the Tax Police anti-corruption unit—an organization independent of the Authority but working with the Authority. The working group became a new bureau in the Authority's organizational structure following the new 2017 legislation.

3.1.3 Supervision on transparency

An important component in the fight against corruption (Galetta, 2014)—and indeed the main aspect, according to Authority President Cantone (ANAC, *Annual Report*, 2015, p. 5)—is the enhancement of transparency mechanisms, following the maxim that "sunlight is the best disinfectant" (Brandeis, 1914), or that "Good government must be seen to be done" (Kierkegaard, 2009).

It is clear that transparency measures operate on different levels and with different aims (Heald, 2006; Merloni, 2008). Above all, in the Italian experience, they perform the function of guaranteeing the rights of citizens affected by administrative action, through the right of access to documents covered by Italian law on administrative proceedings (Law no. 241 of 1990) and thus they operate essentially within the paradigm of "due process" (Galetta, 2014).

Another traditional idea is the democratic and participatory dimension of transparency measures (Carloni, 2014): as such, transparency also binds together participatory policy and the communication activity of public administrations, in which the information provided to the public contributes to a broader involvement of citizens in activities conducted by the authorities, and even the direct management of public interests (and common assets).

Like supervision on anti-corruption, supervision on transparency is performed by the Authority (A) on its own initiative, and (B) upon reporting from the field. Supervision based on its own initiative can be divided into the following:

A.1 Supervision upon request by the Authority Council as follow-up on its own analyses and inspections and reporting from the field;

A.2 Follow-up on parliamentary questions;

A.3 Supervision of the Independent Evaluation Units (OIV), which are a part of all public administration organizations.

We will now discuss these forms of supervision on transparency in more detail.

A.1 Supervision upon request by the Authority Council

In 2015, help on follow-up on inspections was enlisted from the Financial Crimes Police with the following organizations being checked: the Ministry of Industry and Economic Development (MISE), the Municipality of Rome, the Hospital Corporation of Caserta (in the south of the country) and the Municipality of Ancona, on the Adriatic coast. Alongside such issues as discussed in the preceding section on the Municipality of Rome, criticism was also levelled at the transparency section on the organizations' institutional websites. Thus inspection activity was expanded in 2016.

A.2 Follow-up on parliamentary questions

In 2015, three files were opened as a result of questions raised in parliament. The first case involved a refugee centre in Sicily, which was asked to comply with mandated transparency requirements regarding its website.

The other two parliamentary questions were related to local public organizations that manage mountain cable-transport systems in the country's northern Alpine regions. A solution was reached when the organizations published on their websites the management concession agreement from the local political authorities and improved the sites' transparency sections.

A.3 Supervision on the Independent Evaluation Units

Public organizations are subject to specific certifications by their Independent Evaluation Units (OIVs). The Authority therefore checks the websites of a sample of public organizations for the publication of such certifications, both in their form and their substantive content. As a result of field reports, the Authority may interact with the Corruption Prevention Officers of specific organizations. However, the Authority does not have the power to impose specific measures on the public organizations.

In December 2014, the Authority asked the OIVs of all public administration organizations to certify compliance of their organization against a number of obligations. These duties included information about their governing bodies, appointments of top managers and other employees, companies controlled or otherwise owned by their organizations, public tenders and contracts, public works, access to public information and publication of the Three-Year Corruption and Transparency Plans. The following year, the Authority undertook to check 98 public administration organizations, including ministries, independent authorities, regional executives, municipalities of the regions' capitals, and local health care units and other bodies of the National Health Service (SSN).

In conclusion, transparency allows widespread control over the exercise of power and must therefore be pursued, but through generalized disclosure measures that are not dependent on the position of the interested party (Savino, 2013; Gardini, 2014). In this regard, Public Law no. 190 of 2012 provides for a delegation of the regulation regarding communications on institutional websites, which was then implemented by Legislative Decree no. 33 of 2013. This latter decree, which has recently been amended and updated (Legislative Decree no. 97 of 2016), revised and expanded a number of transparency requirements contained in previous legislation. These obligations were expanded after the digital administration code (Legislative Decree no. 82 of 2005) established an obligation for all administrations to have an organized website, in accordance with common principles and standards, with a number of mandatory informative sections (Carloni and Giglioni, 2017). In general terms, the internet has greatly facilitated access to government information (Roberts, 2006) and recent years "have seen trends toward using e-government for greater access to information and for promotion of transparency, accountability, and anti-corruption goals" (Bertot, Jaeger and Grimes, 2010; Cuillier and Piotrowski, 2009) in many countries.

3.1.4 Supervision on procurement

3.1.4.1 General supervision

There has been general supervision of every contract for procurement of works, goods or services. This process is a time consuming one, useful for future reference but ineffective for intervention in real time. For this reason, collaborative and special supervision were initiated in order to intervene in real time, as procurement was being implemented.

3.1.4.2 Collaborative supervision

Collaborative supervision in the public procurement sector was begun in 2015: a preventative measure to guarantee that tender operations and procurement processes progress in the correct manner and to keep misconduct out of the process itself. Requests from public administration organizations for collaborative supervision of their procurement activities came in large numbers soon after rollout of the initiative, and the requests have only increased over time. This seems to point to overall shortcomings in the public organizations' procurement processes.

Collaborative supervision is regulated by ad hoc protocols signed between the Authority and the individual procurement units, which include the following information: the specific case in point to be overseen, the list of documents to be reviewed by the Authority, a description of the supervision procedure, the time-frame of collaboration between the Authority and the procurement unit, possible interim reviews to render the supervision more effective, and the possible lawfulness clauses to be included in the tender notice. It is not possible to ask for collaborative supervision when the call is already under way. Protocols list the specific documents that are subject to collaborative supervision: the decision to launch a procurement process, the tender notice, public letters of invitation to bid and the terms of the contract. Information is also included about the execution of procurement: appraisal of modifications, settlement, amendments to contracts and annulment of contracts.

Collaborative supervision follows a specific procedure whereby documents are sent to the Authority in each phase of the process, then two interaction phases take place between the procurement unit and the Authority. In the first phase, the Authority sends the procurement unit its observations on the documents. The procurement unit then replies with a response. The Authority then sends a "closing" memo. By following this procedure, the Authority makes its observations in a timely manner; the unit is free to accept them and remains responsible for all its further action.

3.1.4.2 Special supervision

The Authority also undertakes supervision specific to four categories of cases, which is known as "specific" supervision. The four categories are:

1   Large public works, dealing mostly with awards of contracts to general contractors and with execution of contracts;
2   The energy and waste management sectors, dealing mostly with fragmentation in the way procurement is conducted and with extension of contracts, often awarded by public administration organizational units, e.g. municipalities, whose territorial scope (*ambito territoriale*) is usually much smaller than would be conducive to effective management;
3   Seaports and airports, dealing with competition and transparency in the process of awarding concessions to a specific managing company;
4   Further specific investigations carried out on certain large contracting authorities (e.g. the Municipality of Rome, the Rome local transport company [ATAC] and the Italian Postal Service).

Special supervision took place in 2015 on two major events: the Milan 2015 Expo, a global fair, and the Rome Extraordinary Jubilee of Mercy, a religious event. The Milan Expo supervision began in 2014 when a special operations unit was created by a Law Decree (see Chapter 8), and then in August 2015 the Presidency of the Council of Ministers, on the basis of the same decree, decided the Authority should also perform preventative legitimacy checks on procurement for the Jubilee. The Milan Expo special supervision can be considered the first of its kind, after which the tool has been systematically used for large events, initiatives and works of national or strategic interest as a means of guaranteeing transparency, due process and quality of administrative decision-making from the outset.

This tool marks a cultural change: the Authority no longer intervenes to sanction and condemn illicit behaviour ex post, when the damage done is often difficult to remedy, but instead aims to prevent anomalies *ex ante* by guiding the administration towards better and more transparent choices and discouraging inappropriate operators from responding to calls to tender.

3.1.5 Supervision of the certification system

The Authority oversees the certification system for private companies within the public works sector. Crucial to the system are the certification bodies, overseen by the Authority, which validate that private companies looking to enter public tenders and perform public works have are suitably qualified. This system undoubtedly simplifies the award of contracts because procurement units only need to consult the list of certified companies rather than checking all the tendering companies themselves. A further benefit is that the requirements are independent of specific tenders, which appears to be a useful limitation for the procurement units: because they cannot themselves define tendering requirements, full participation in the tender is allowed. Nonetheless, there are issues. A key issue is the certification system's conservative approach, which limits opportunities for new entrants: the certification system as it stands is based on a historical record of companies, both as regards actual projects executed and historical costs. As such, the system has, on the one hand, afforded objective verification and certification of company requirements but, on the other hand, it has "frozen" the market on the presumption that the operating capability of a company is based on its past performance rather than on its present capabilities. Such barriers to entry were raised even higher when the time period for measuring prior performance and accounts was extended to the "last decade".

The new code of procurement is intended to address this, which is a future task for the Authority. Meanwhile, the Authority oversees the qualification system as it stands, an activity performed both upstream and downstream relative to the certification bodies. The first function is about the bodies' compliance with requirements such as independence, ethics and sufficient capital. The second is about checking the bodies' actual certifications.

3.2 Regulatory activities

In the following sections we will describe activities related to the Three-Year Plans on Corruption and on Transparency, on the Authority's acts of interpretation, and on standard prices.

3.2.1 Analysis of the Three-Year Plans to Prevent Corruption

As discussed in Chapter 2, in 2015 the Authority analysed 1,911 Three-Year Plans to Prevent Corruption, which corresponds to the number of public administration and wider public-sector organizations. It looked at the 2015–2017 plans as published on the organizations' institutional websites; if 2015–2017 plans were not available, 2014–2016 or 2013–2015 plans were examined. It should be mentioned that this was the first time these plans had been examined, with the Authority undertaking this operation for the first time since Public Law no. 190 of 2012 had mandated the creation of the plans.

The Authority first checked that plans had actually been written and approved by their own organization, and then checked their quality, looking for possible relevant issues and the implementation of a corruption-prevention strategy. The evaluation focused on the risk management process, the planning of preventative measures and on the integration of anti-corruption measures with other planning instruments and documents.

3.2.2 Regulation on corruption prevention

This bureau's main task is the National Anti-corruption Plan but among its other activities is regulation on corruption prevention in the following areas:

1 Guidelines on whistleblowing;
2 Guidelines on SOEs (i.e. entities controlled by or simply those in which public administrations participate [i.e. in which they have a minority share]).

3.2.2.1 Guidelines on whistleblowing

A legislative decree of 2001 for the first time institutionalized whistleblower protection in Italian law (amended and improved in November 2017), with the aim of encouraging public administration employees to report corruption. The law protects whistleblowers' identities and also protects them from discrimination and possible sanctions. Whistleblower reports are to be submitted to their line managers, or to the judiciary or to the Court of Auditors, a special branch of the judiciary for audit in public administration.

In 2015 the Authority published specific implementation guidelines on whistleblower protection to encourage public administrations to actually address this subject, a full 14 years after the decree. This was done following prior consultation.

The Authority's guidelines define the boundaries within which organizations fall that are subject to this norm and the privacy measures that public administration organizations need to put in place to protect whistleblowers. The Authority produced a proposal for an automated software system to manage whistleblower reports which it offered free to public administrations to save them the expense of buying software on the open market. The platform is called "open whistleblowing" and the prototype allows for both anonymous and credited reports. It also allows internal exchanges between functionaries responsible for the processing of the reports. Whistleblowers can check the status of their report and can communicate with those who are processing it (following the 2017 law).

3.2.2.2 Guidelines on SOEs

The Authority and the Ministry of Economy and Finance (MEF) developed a joint policy regarding corporations and other bodies under public control. In general, the policy concerns bodies regulated by private law that are under complete or partial control of public administration organizations. In 2014, the Ministry and the Authority jointly approved a general statement and the Ministry also published a directive with guidelines about corporations under their own control or with which they participate.

In 2015, the Authority published new guidelines on how private law organizations under public control should implement anti-corruption and transparency measures. This also went through a prior consultation process which engaged many of the stakeholders: local entities, corporations, foundations and associations. A big distinction was introduced here: corporations and other bodies under public control will undertake anti-corruption and transparency activities very much like public administration organizations, but corporations and other bodies in which public administration organizations only participate (minority share) will undertake anti-corruption and transparency only on their public-interest activities (see also Resolution no. 1134 of 8 November 2017).

3.2.3 Regulation on transparency

The Authority's statute makes it clear that anti-corruption and transparency activities are indivisible. Transparency concerns access to information about the structure, organization and activities of public administration. The connection between transparency and anti-corruption is also promoted by international organizations.

In 2015, the Authority defined regulation in three areas:

1   Guidelines on corporations under public control or partially owned by
     public administration (simple public shareholding);
2   Guidelines on restrictions on spending (with regard to corporations and
     overseen bodies in general) to the benefit of the overseeing bodies;
3   A resolution on the exercise of powers of sanction.

3.2.3.1 Guidelines on corporations under public control or partially owned by public administration

These have been dealt with in the previous subsection.

3.2.3.2 Guidelines on restrictions on spending (with regard to corporations and overseen bodies in general) to the benefit of the overseeing bodies

A law mandate applicable to all public administration organizations requires
them to publish the names of the bodies they supervise, those they fund and
those with top managers appointed by public organizations. They also need
to publish the names of corporations and entities they control or in which
they participate.

Specific data must accompany the list of such entities: name, percentage of
share owned, time-frame of engagement, annual cost, number of representatives on the governing bodies of the controlled organization and their remuneration, financial reports from the last three years, and finally the top
manager's name and remuneration.

3.2.3.3 A resolution on the exercise of powers of sanction

In 2015 a clarification was issued about the Authority's power of sanction in
cases of violations of transparency regulations and new legislation. Before this
new delegated Legislative Decree no. 97 of 2016 was issued, the Authority
would denounce the violation, impose a reduced sanction and send the case
to the Prefect, who had full authority to sanction. Following the introduction
of the new legislation, once the Authority has denounced the violation it
now has full power to sanction.

3.2.4 The Authority's interpretation of the law

The Anti-corruption Authority regulates public contracts, ensuring the law is
properly implemented. But the legislation here is quite complex. The Authority's mission is, strictly speaking, to oversee public contracts, both in award and
in execution phases, and its objective is the efficiency and effectiveness of such

contracts, through competition and transparency. The Authority's regulatory function is intimately connected to its supervision function, especially "preventative" supervision, the latter being carried out through the interpretation and integration of relevant laws. The Authority's work in this regard revolves around providing managerial models to public administrations' procurement units.

The Authority's interpretations of the law are presented through their decisions, guidelines and templates for calls for public bids. These outputs are published following a process of prior consultation, or "notice and comment", thereby offering both procurement units and companies (the stakeholders) the opportunity to voice their requirements and have them taken into account.

Many of the Authority's regulatory activities undergo *ex ante* "impact analysis of regulation" and ex post "impact assessment of regulation". The former offers explanations for the non-acceptance of any of the prior consultation's proposals, whereas the latter aims to learn whether or not the regulation was implemented and, if it was, its effectiveness.

In 2015, the Authority published nine decisions regarding public contracts under the following headings:

1   A new discipline concerning interim relief;
2   Awards of architectural and engineering services contracts;
3   Companies in default or in receivership;
4   Awards of maintenance service contracts for facilities;
5   Award of security services contracts;
6   Project financing;
7   Consolidation of procurement by municipalities;
8   Awards of services contracts to non-profit organizations;
9   A standard template for calls for tender of goods and services.

3.2.5 Benchmark prices and standard costs

The Authority's activities regarding contracts and prices are based on a "national database of contracts" which the Authority inherited from one of its constituent organizations, the Authority on Public Contracts. In 2015, the Authority worked to identify benchmark prices and standard costs to be used in public procurement, both in the health care sector and in the non-health-care sector, which will allow for spending reviews and identification of corruption. However, this is a complex undertaking considering the potential diversity of services and goods that must be rendered comparable. To this end, the Authority collaborated with the Italian National Institute of Statistics to develop an econometric system which allows price comparisons even in the most complex cases. Satisfactory experimental results were obtained for cleaning services in the health care sector.

This brings to a close our description of the Authority's activities. We will now turn to the Authority's organizational structure, examining the personnel and their roles in order to gain an understanding of where the Authority is heading in the next few years and what will be its priorities.

3.3 The Anti-corruption Authority's organizational reality

3.3.1 Introduction: strategy and structure

In this section, we will discuss the full range of functions across the Authority's organizational units. Thus far, Authority activities have been presented in a somewhat aseptic manner, but we need now to observe the human angle and examine it from an organizational perspective. The Authority does not operate in a vacuum and remains a part of public administration; it is therefore subject, like other administrations, to the momentum of its own history and it must abide by its own budget. It has its own staff and, although laws mandate its functions in the abstract, the reality of the implementation of law on the ground is greatly influenced by incumbent staff and their competencies. We will therefore attempt to gauge the relative significance of the Authority's activities in the full awareness that structure and strategy interact and affect each other (Chandler, 1990). We will look at the structure as a whole and make some observations about the nature of the Authority, based on the resources committed to each of its areas of activity. The starting hypothesis is the Authority has three main substantive areas of activity: anti-corruption, transparency and supervision of contracts. It would appear that contract supervision has become dominant, especially in light of its new tasks.

In the following we will describe some of the tasks assigned to the Authority under the new contracts code and a possible consequent issue concerning the Authority's future character. We will go on to describe other relevant factors, such as the Authority's organizational heritage and its budget restrictions. We will close this section with a proposed future strategy for the Authority in light of its new tasks.

3.3.2 The new contracts code

In 2016 a new Public Law on public procurement was approved which gave the Authority additional functions: the Authority now manages not only the qualification of certification bodies, but also a new register of the members of bid evaluation committees as well as a new register of companies which keeps track of each company's performance in previous contracts and evaluates them, awarding a "company rating" based on prior technical performance. It should be noted here that there is also a "lawfulness rating" for companies—a completely different rating—which was entrusted a few years ago to the Antitrust

Authority, an entirely separate body. Activities concerning public works contracts therefore come to predominate within the Authority, as each of these new tasks requires dozens of functionaries.

3.3.3 A possible issue

As a result of these new responsibilities, in November 2016 the Authority decided on a new organizational structure, to be implemented over time, involving 19 line units. Of these 19, 15 are devoted to contracts supervision, three to anti-corruption and one to transparency. The units tend to have the same weight in terms of personnel and budget.

These measures appear at odds with the organizational chart we have been describing. In fact, at present the Authority is structured along the two key lines of supervision and regulation, a structure that cross-cuts vis-à-vis the three substantive areas of anti-corruption, transparency and public procurement in order to strike an equilibrium among the different areas. This equilibrium is at risk, though, owing to the new legislation on contracts, which may call for more resources for public procurement.

A possible issue arising from this is that such structural changes might be intimations of a specialization by the Authority on public works with a concomitant downplaying of its initial mission as an across-the-board, nationwide and administration-wide agency with competence on corruption and transparency. Legislative activities almost by definition take place in well-known territories, lag behind knowledge and answer the logic of political action. There is therefore a real risk of the Authority being scaled back to a sectoral authority working mostly on public procurement. It is worthwhile to notice that while the new procurement code was being discussed in parliament, the new Legislative Decree no. 97 of 2016, on transparency, was being discussed at the same time, but this new transparency legislation did not lead to much of an increase in the Authority's functions. It appears a choice had to be made about what was essential and what could be relinquished: the Authority could not "do it all".

3.3.4 Organizational heritage

This risk we have identified is further exacerbated by the fact that the Authority's organizational heritage is in the very area of contract supervision. As discussed, its structure derives from the merger of two former bodies: the Commission on Evaluation, Integrity and Transparency of Public Administration (CIVIT), working under the aegis of the Minister for Public Administration, and the Authority for the Supervision of Public Contracts (AVCP), working under the aegis of the Ministry of Infrastructure and Public Works. As described earlier in this book, the number of managers was reduced through attrition and their salaries were also reduced following the merger.

Nonetheless, it is clear that this public works heritage made it hard to balance the Authority's activities across the whole public administration spectrum, requiring as it did a better distribution of staff competences for its three key areas of anti-corruption, transparency and public contracts supervision. However, new powers to manage its own organization and personnel were granted in April 2017.

If we want to understand where the Authority is heading, it is important here to consider "organization" in its broader sense. Organization consists of seven elements, according to Pascale's "7S" model (Pascale and Athos, 1981): strategy, structure, systems, staff, skills, style and shared values. The "staff" element is critical here: the organization is inevitably affected by the demographics of its own personnel—the knowledge they share and their culture. Therefore, if among a total staff of 350 six out of seven have a public works orientation, there is a challenge in adequately delivering other, very different functions, such as transparency and anti-corruption; meanwhile, because of its heritage, the Authority places undue weight on contracts supervision.

3.3.5 Budget restrictions

Budgetary considerations should not be ignored: the Authority's capability is also a function of its finances and the financial latitude it has to manage its own structure and strategy. With an initial budget of around €65 million, the Authority was required to save 20% of this year on year, thereby reducing it by €13 million to €52 million. Under such a tight budgetary regime, the Authority has to confine its activities to mandatory statutes with little scope for new ideas that need to be tested; it engages only in "continuous organization of official functions, and specified spheres of competence" (M. Weber quoted in Merton, Gray, Hockey and Selvin, 1952, p. 19). The Authority runs the risk of doing just what external parties want it to do, be they parliament or the lobbyists.

In short, budget constraints have an adverse effect on the specific activities of the Authority's mission, by which we refer particularly to its anticipated innovations in the public administration system. These are ipso facto not defined in law because it was the Authority's task to formulate them: how to tackle corruption and what specific organizational strategies can be devised in pursuit of this goal. However, from 2017, the budget is increased by €10 million to €62 million.

3.3.6 A possible strategy

One direction in which the issues outlined above could unfold is that the Authority apprehends this as an opportunity to promote anti-corruption from within the public works domain itself. The Authority could acknowledge public works as a key area of anti-corruption activity, thereby reconciling its organizational imbalance with its anti-corruption mission. Going further: maybe such sectoral emphasis could produce a domino effect and bring about a general call for effectiveness in the other sectors of government, not only as regards expenditure but also in delivery of services, in recruitment and so forth. Furthermore, there are important issues to be addressed in the public works domain itself, such as opening up the Italian market to foreign operators.

3.4 Conclusion

In this chapter we have tried to describe systematically the activities of the bureau of the Italian Anti-corruption Authority. We have noted the vast array of measures in which the Authority intervenes. We note also the "meta" nature of the Authority's activities, i.e. it does not work on corruption per se, but on those factors that may induce corruption and those that may reduce it. Its strategy is to remain remote from operations and to be present only in an indirect manner. To this end, it relies on the internet with many of its inter-actions with public administration taking place online, a manifestation of the distance the Authority establishes from the control of public organizations.

It is no easy task pursuing its prevention strategy, no less so than if it were adopting a repression approach. The Authority aims to make organizations responsible for themselves, rather than "doing" anti-corruption on their behalf; in other words, it does not tell organizations what to do but ensures they do something on their own initiative. Metaphorically speaking, the Authority is all about the rules and it tries to stay out of the game as much as possible. Nonetheless, we should appreciate that anti-corruption activities are very much intertwined with day-to-day activities and as such the Authority's work might sometimes be quite similar to regular administration. The risk is that the Authority's regulations about minimum standards will be interpreted by addressee organizations as objectives, as targets of "satisficing" behaviour.

Notes

1 Pre-litigation (*pre-contenzioso*) concerns counsel (*parere*). The Authority advises on matters of possible litigation between tendering units (*stazione appaltante SA*) and economic units (companies [*operatori economici*]), who may ask the Authority for counsel before going to court. In 2015, there were 940 such instances of counsel. Apparently, it often obviates litigation.
2 According to the Cambridge Dictionary: "lack of care, judgment, or honesty in the management of something".

Chapter 4

Country comparison of anti-corruption efforts

In the first three chapters, anti-corruption efforts have been illustrated with different levels of detail and in different countries. We will now draw some comparisons among the diverse evidence that has been presented and offer some observations about the countries' institutional and organizational landscape.

In order to offer a systematic comparison, this section will be laid out according to the parameters that were presented in Chapter 1 in describing the individual countries' approaches, namely the institutional setup; the scope of the effort; the emphasis on public vs. private corruption; prevention vs. repression; relationship with the field vs. concentration on a few issues; objective vs. subjective measures; and the relationship with transparency concerns.

4.1 The institutional setup

The institutional setup is about the positioning of the anti-corruption effort within the constitutional framework of the government. The basic question is whether the anti-corruption organization is independent from the classical three branches of power or located within one of them, commonly the executive branch.

4.1.1 Independence from the branches of power

The key factor in the institutional setup is the independence of the anti-corruption organization from the branches of power, or its organizational position and constitutional status.

France's Agence Française Anticorruption (AFA) is integrated within the executive. Although it has a magistrate and its own budget, it is under the control of France's Ministries of Budget and Justice. In contrast, Italy's Autorità Nazionale Anticorruzione (ANAC) can be described as an independent authority. Board members are appointed by the executive every six years and this appointment is ratified by parliament. Like France's AFA, Italy's

Authority reports on its activity to parliament annually. In the UK, the key institutions are the Serious Fraud Office (SFO) and the Joint Anti-corruption Unit. The latter is under the control of the Prime Minister while the SFO can be considered a semi-independent non-ministerial government department under the oversight of the Prosecutor General. The USA does not have an independent authority, with the Office of Government Ethics structured within the executive; however, appointment of the head of the Office is ratified by parliament (Congress) and the Office reports to parliament. It is worth remembering here that the US anti-corruption effort was started as long ago as 1978, long before the 2003 UNCAC. In the case of Australia, using the example of the New South Wales (NSW) ICAC, we can attest that this authority is independent, with its own budget and autonomy from the executive and legislative powers.

When we talk of the independence of the anti-corruption effort, we mean independence from the executive. However, upon closer investigation, the importance of this issue seems to recede. The daily workings of anti-corruption organizations are so pervasive and full of micro-activities that political or bureaucratic influence does not seem to have much of an effect. Furthermore, where major issues are concerned, formal independence from the branches of power will not prevent a Prime Minister or a member of parliament from attempting to meddle with anti-corruption matters.

The UK's approach seems to be the lightest: only one small unit, and within the executive. It is a decentralized approach, with various institutions delivering piecemeal anti-corruption measures, each in their own domain. The UK's lack of independence from the executive branch should give us pause to consider the possible advantages of such a circumstance, which need not be ruled out a priori.

Formal independence from the branches of power, however, does mean that political changes take time to affect the anti-corruption effort and an authority appointed by one party will not be immediately dissolved when its opponents come into power. This can ensure continuity of the anti-corruption effort and the medium-term implementation of plans. But in politics and the public sector, of course, the future is never certain, and a new governing party may still make sweeping changes. At a minimum, independence is one among those organizational arrangements that ensure incremental change among top-level management, which—like rotation—provide for continuity and mitigation of animosity.

However, the UK seems to have a very different attitude to the importance of continuity: it appoints a new Champion of Anti-corruption annually, the theory being that new blood is better for performance than continuity and that institutions are sufficiently resilient that a change of top management does not mean they have to start over.

4.1.2 The shape of organizational arrangements

As anticipated, we can observe a recurrent theme among the different anti-corruption organizations: a central agency with a functional (non-hierarchical) relationship with the officials within the supervised agencies. This is not unsurprising, although it is not the only possible arrangement. For instance, in Italy the idea was put forward that the field officers responsible for corruption prevention should report hierarchically to the central agency. At the time, this appeared to exist only as an opposing point of view, expressing a very centralistic model. However, it could be said that France's 70 AFA agents represent an organizational structure not too far removed from the Italian proposal.

The UK presents a very different picture: its anti-corruption effort does not take the form of one specialized organization tackling one specific problem, as with all the other cases we have observed. Other countries' model is that of a one-to-one relationship between a problem (corruption) and an organization to address it, which is the common approach in public administration and in organizations in general. However, this need not always be the case, as the UK proves.

The "one problem, one organization" approach implies an underlying hypothesis that the new organization's interaction with previously existing organizations will have no systemic effects. Organizations are not necessarily "linearly additive", i.e. if a new problem arises, a new organization will be set up to address it and the outcome will be the sum total of the organizations' capacities. This fallacy belongs in the original rational paradigm of organizational behaviour that has its roots in the Weberian legal rational model of authority.

Australia is, as a country, by definition decentralized to begin with, dealing with corruption piecemeal in a country where corruption is seen as a problem to be dealt with as a federal offence. However, we need to be aware of relative sizes. Australia has a population of about 25 million; dividing the anti-corruption effort equally among five state institutions means each institution operates within a population of about 5 million on average—one-tenth of their European counterparts examined in this study. The importance of this should not be ignored. Politics needs a minimum scale, as James Madison pointed out,[1] to avoid too much personalization of issues; public administration, however, being a mostly white-collar, bureau operation, does suffer from diseconomies and dysfunctionalities of scale.

In contrast, the US is about five times bigger than its European counterparts in this study. It has confined its own anti-corruption effort to its federal government for constitutional reasons, although the policy might be sound from a scale point of view as well.

4.1.3 Modi operandi

The French APA supports a prevention function via the publication of an annual Strategic Plan; and its key functions are centralizing and analysing data on corruption and providing advice to other authorities, judicial authorities and public institutions on corruption-related matters. The body does not pursue judicial cases but analyses systems of corruption utilizing its 70 agents.

The Italian Authority promotes an explicit cycle of planning and feedback among its overseen organizations, requiring public administrations annually to publish an update to their Three-Year Plans to Prevent Corruption and for Transparency and Integrity.

As stated, in the UK there is no one central authority tasked with preventing corruption. Every year the UK government prepares an anti-corruption plan, which attracts criticism for being merely a statement of intent rather than a comprehensive programme of action.

The US Office of Government Ethics provides direct support via desk specialists to the 4,500 officials in the Inspectors General offices (or "Agency Ethics Officers" as per the definition in the Standards of Ethical Conduct for Employees of the Executive Branch). They also undertake training as well as "institution building", "aggregating" and support activities.

The NSW ICAC, in a cyclical planning activity, defines a strategy and publishes a report every two years which includes performance indicators and measures actual results against prior objectives.

It is interesting to compare the role of Italian Corruption Prevention Officer with that of the Inspector General in the US public administration. The former is a facilitator and a lone figure with no staff, whereas the latter is an investigator—the 130 Inspectors General have a total staff of 4,500.

We have noted that the organizations reviewed all have a nationwide scope, vis-à-vis sectoral- and organization-specific efforts, something that can be found discussed in the literature. For instance, Klitgaard (1988) reports on the anti-corruption effort in the tax administration of the Philippines. However, some organizations only work within central government (as in the US case) while others include central, regional and local government (as in the case of Italy).

Even though there may be a specific organization in charge of anti-corruption, we notice that some tasks are nonetheless entrusted to other organizations. This is generally because of overlap with other organizations' tasks and historical factors related to institution building. As a result, we find ourselves observing a variety of organizations in each country. The anti-corruption effort, then, as is the case with many others functions of government, is a result of multiple efforts.

On a different note, the plans should be understood as instruments to make individual organizations responsible, to generate measures of output and outcome. Intermediate measures are also critical: in the face of the difficulty of measuring corruption per se, some institutions resort to *measuring their own output*—measuring "containment action" (*indicatori di contrasto*, in the words of the Italian Authority) as a proxy for their own effectiveness in the field. This needs further analysis.

On a different note still, all anti-corruption organizations seem to have a pathway for legislative reform. They can all make proposals about the tightening of their own legislative mandate and about the general mandates of the organizations they oversee.

4.2 The power and scope of the organization

This dimension is about the "perimeter" of the anti-corruption effort. That could involve the country's public administration and not its state-owned enterprises, or only a segment of public administration, e.g. only federal but not state, regional or local public administration. Also, it could apply to all business firms and especially the large ones which may have important dealings with foreign governments and thus be at risk of foreign corrupt practices. Corruption of private parties in dealings with other private parties can indeed take place, but it is probably dealt with as industrial malfeasance and appears not to be very much on the public authorities' radar.

4.2.1 Differences in "perimeter"

France's AFA, in late 2017, was requested to implement a programme to prevent and detect corruption in state-owned companies with at least 500 employees and a turnover exceeding €100 million. It was subsequently called to supervise the whole public administration of France, almost 5.4 million employees.

The Italian Anti-corruption Authority and the nearly 11,000 Corruption Prevention Officers in Italy oversee 3.3 million employees of the executive branch and regional and local government. The total rises to 4 million once SOEs are taken into account.

The UK SFO has the power to investigate all the country's public sector, almost 5.4 million people. They also have power to investigate the private sector.

The US Office of Government Ethics and the 130 Inspectors General supervise 2.7 million civilian employees of the federal government.

In Australia, the NSW ICAC has approximately 110 employees and oversees about 300,000 public employees.

The definition of corruption changes according to the perimeter of action of each of our authorities or institutions under study. In fact, the total scope of the anti-corruption effort in the USA appears to be relatively limited compared to Italy. In fact, US public administration employs about 24 million people, yet only 2.7 million are under the oversight of the Office of Government Ethics. One may observe that this is one way to manage corruption—by transferring it to the private sector. Things that have the same name may be very different in reality. For instance, health care units have about 400,000 public employees in Italy: they are subject to anti-corruption oversight by the anti-corruption authority, and corruption in their ranks is perceived as public corruption. The same may not be true in the US, where health care is organized through private parties and health care personnel are not public administration personnel, so it is possible that the *international* perception of corruption in this sector may not link it with US public administration. This is interesting because it may point towards organizational measures involving the private sector to contain corruption, which are especially important when a preventative approach is being taken.

4.2.2 Intergovernmental relations

The cases of federal countries such as the USA and Australia draw our attention to intergovernmental relations. Do anti-corruption efforts cut across the different levels of government? The answer, in the case of France, Italy and the UK, is "yes" (although degrees of autonomy are likely to be in place for England, Scotland, Wales and Northern Ireland); yet a resounding "no" for the USA and Australia. What are the pros and cons of such arrangements? Benchmarking between the different authorities in Australia seems to offer a potential benefit.

On the one hand, in the case of Italy the Authority has a mandate to supervise central and local government. It is quite similar in this regard to the UK's SFO which can investigate public and private administration and companies at central and local level.

On the other hand, the US Office of Government Ethics has authority only at federal level. As such, US federal legislation does not offer any indication of what could or should be done in the individual states.

Ultimately, in Australia the NSW ICAC is able only to investigate at state level and there is no central authority at the federal level with a mandate to tackle corruption.

4.3 Definition of corruption: public vs. private

France's AFA has a mandate to oversee state-owned companies and has subsequently acquired authority over public administration. It has no competence to address corruption in the private sector. In Italy, the Authority is tasked with tackling corruption in the public sector, which includes foundations and state-owned companies. There is no authority addressing private corruption. In a contrasting arrangement, where the boundaries between public and private are not as clear, the UK's SFO has power of investigation of both public and private corruption. The US Office of Government Ethics is only concerned with corruption in the US federal public administration, while concern with private corruption is left to the Department of Justice and the enforcement of the Foreign Corrupt Practices Act. In Australia, the NSW ICAC is concerned with public corruption.

Opinion polls about corruption in any given country tend to indicate that public opinion sees corruption as an issue related to that country's public sphere rather than to its private sector. Considering that private corruption comes under more scrutiny for its foreign corrupt practices (with foreign public sectors), one might add that public corruption is thought of as domestic corruption while private corruption is more about international corruption.

Regarding the distinction between public and private corruption, we may recall here the corporate social responsibility (CSR) concept and movement: a complex range of ideas and practices that have been implemented by both large businesses and non-profit organizations. The movement involves academic studies, business practice, government regulation and incentives, and it has as its global apogee the UN Global Compact (UNGC).

The practice of CSR, although much criticized, has nevertheless been widely disseminated across the globe; it enjoys waves of fashion and then goes quiet for a while but it is highly unlikely to disappear.

The strongest connection between public corruption and CSR is to do with supply chain responsibility, because all actors have this issue in common and it has become an acknowledged area of social concern for private organizations. After all, public administration procurement is supply chain management, something that is never sufficiently scrutinized, either in the private or the public sector. In this context, the kind of corruption we are concerned with takes place between a public party and a private one. We therefore find ourselves looking at the same phenomenon from two different perspectives. On one side is public corruption and the UN Convention Against Corruption (UNCAC); on the other is private corruption and the UN Global Compact (UNGC). We need to leverage this connection.

An important consideration is when corruption is defined as maladministration. A preventative approach is thereby implied because the issues being dealt with are not criminal activities. It is also a very ambitious approach because it implies a fundamental aim of improving the overall performance of public administration. The task is a proactive one, not one of simply preventing decline.

4.4 Mission: prevention or repression of corruption

The prevention approach is comprehensive by nature as it includes—indeed fosters—organizational and legislative change. An openness to organizational change could be leveraged via this cross-country comparative analysis because, as we have seen, organizational measures to tackle corruption may include new economic models for entire sectors.

The preventative approach is common in France's AFA, Italy's Authority and the NSW ICAC, whereas the mission of the UK's SFO and the US Office of Government Ethics is oriented towards investigation and repression.

It may be observed that all organizations perform repression to some degree. Even those most configured towards prevention (such as France's and Italy's) conduct investigations and report cases to the judiciary. Likewise, all organizations perform prevention to some degree. The US Office of Government Ethics mostly supports the Inspectors General in their preventative functions, even though the latter appear to work mostly on repression.

4.5 Operations: (micro-)reporting from the field or concentration on big issues

In its first year, as mentioned previously, France's AFA focused its activities on only the largest SOEs, whereas the Italian Authority had to manage the impact of intensive reporting from the field within its own structure, as input was coming both from investigations taken on its own initiative and from whistle-blower public-sector employees and citizens. The UK's SFO investigated only a limited number of cases. The US Office of Government Ethics does not interact with the field: it does not deal with citizens or individual employees of public administration but only with the Inspectors General offices. In Australia, the NSW ICAC received 2,489 reports in 2016–2017, combining different types of cases, and also answered 105 requests for corruption-prevention advice.

Concentrating on a few issues may allow the organization to "pick the low-hanging fruit", which is standard advice in anti-corruption practice (Klit-gaard, 1988). Delivering quick initial results is deemed an appropriate way to generate consensus around the anti-corruption effort.

However, it is almost inevitable that an organization will ultimately turn its attention to field operations and address cases at various orders of magnitude. Even those anti-corruption organizations that concentrate on the big issues—like the UK SFO or the NSW ICAC—still need to sift through a large number of reports in order to select the few cases they want to pursue in depth.

In fact, receiving, indeed encouraging, reports from the field—and responding to them—helps to create a culture of non-corruption because it shows citizens—and public employees and private businesses as well—that "there are still judges in Berlin".[2] However, large operations may take a toll on the organization if the workload is not managed in a sustainable way, and the project may backfire if the organization ultimately breaks down.

4.6 Focus of action: objective vs. subjective measures

Plans are objective measures. Everybody does central reporting. Only France and Italy appear to be mandating individual organizations to draw up their own plans. Such plans loom large in an anti-corruption strategy because they involve every single organization and impose a task on each one. Such measures attract criticism from public organizations, although the very fact that they find things to complain about may imply that at least they are engaging with the project.

Last but not least, such wide-scale measures generate a large number of documents that need to go through a feedback process; however, the anti-corruption organizations are only able to deal with samples and there is as yet neither the public awareness nor the capability to provide the necessary feedback in a decentralized and networked process. Anti-corruption is yet to go viral online. We need to work on this.

Subjective measures relate to individuals' impartiality and conflicts of interest. Such measures might also be enforced through codes of conduct, for instance. Subjective measures have the capacity to create a class of dedicated and trustworthy functionaries, something the International Monetary Fund (IMF, 2016) affirmed the need for.

Anti-corruption bodies addressing the personal wealth of politicians and high-level public employees are working in this domain, and all the countries examined have to some extent a provision to keep track of personal wealth. However, training within public administration organizations appears to be addressing these subjective issues in a broader fashion, reaching deeper into organizations with the goal of changing individuals' behaviour and instilling a personal drive towards integrity and impartiality. The Australian commissions are working along these lines, as is the US Office of Government Ethics. Our analysis of the Italian case reveals a strategy underlying both subjective and objective measures where the two types of measure reinforce one another. On the one hand, subjective measures are meant to foster an organizational climate whereby it is more difficult to act in a corrupt fashion, because of individuals' awareness and peer pressure; on the other hand, objective measures "dry out" the terrain for corrupt behaviour. Objective measures, as already stated, include the identification of at-risk situations and organizational positions—such as the head of a procurement department—and shining a spotlight on them. They also include the implementation of rules connected with these risks, such as personnel rotation.

4.7 Relationship with transparency

France has a dual system: transparency issues rest with the Commission d'Accès aux Documents Administratifs rather than being entrusted to the anti-corruption organization. In the US, transparency rests with the executive, in the Department of Justice. In Australia, anti-corruption is a state matter whereas transparency is federal. Four out of five of the countries analysed show only an indirect interest in transparency in the public sector. Sometimes this is entrusted to a privacy organization (like in France) or a different part of government. The Italian Anti-corruption Authority has control over transparency because freedom of information rules came into existence in this country only after the anti-corruption organization had been established; it appears to be the only anti-corruption organization with authority on transparency.

4.8 The anti-corruption plan as an instrument of engagement

Government by example is key because the "everybody's doing it" mentality is universal. Picking low-hanging fruit sets an example for the rest of the community because it engages every citizen. It is useful here to highlight an outcome of our cross-country analysis: the anti-corruption report as a management tool. This is rather new and might be described as an attempt at "documenting irresponsibility". The anti-corruption plan is a document that in all its essential parts (think of risk analysis, for instance) is basically a CSR report, a well-known instrument in the private sector. It is a way of engaging every organization and making them aware of their own responsibilities—and a key instrument in France and Italy. So the French compliance programmes and the Italian Three-Year Plans to Prevent Corruption are basically CSR reports within public administration. This is a rare instance of government management (rather than policy) accountability in action.

The plan is an instrument that engages and discloses. It is a declaration by the organization, reluctant though it may be. The plan may tell lies, like a psychiatrist's patient may tell lies, but, if the analyst has a sound understanding of the public-organization brain, she or he will be able to deduce the health of the organization. It is not surprising that the CSR movement—the private sector's counterpart to anti-corruption—has placed so much emphasis on reporting.

It might be observed in this regard that the most recent manifestations of the UNCAC are technically very similar to manifestations of the UNGC. The UNGC anti-corruption principle, #10, focuses on—indeed calls for nothing but—business responsibility, just like all the foregoing Compact principles. Coming from this, the Global Reporting Initiative (GRI) is one of many initiatives intended to help put the UNGC into practice. There is a parallel here with the anti-corruption planning and reporting instrument. An anti-corruption plan includes risk analysis which in turn implies (but as yet does not require) connection with both budget and HR functions. An anti-corruption plan also stipulates the implementation of a code of conduct, another staple of private-sector CSR reporting.

4.9 Summary of Part I

So far we have looked at public administration's efforts to tackle corruption, examining the active parts of public administration: the anti-corruption organizations. We have gained some insight from a comparison of the experiences of different countries and from an in-depth investigation of one particular country. In doing this, other issues have surfaced, and they are as follows.

There is a general question mark hovering over all anti-corruption efforts: how effective are they? The state of the art in evaluation of the impact of anti-corruption measures consists of specific studies, with no systematic statistical value consistently measured over time. In this respect, all we have to go on are opinion-based surveys. Therefore, for any serious attempt at evaluation, we are left with the specific studies. Among these a critical voice is heard:

> A final accountability mechanism is the cadre of inspectors general, who now hold offices within most federal agencies, including the Department of Justice. Inspectors general have the power to investigate legal violations, sometimes including crimes, within the executive branch. Some can be discharged by the agency head, but some can be discharged only by the president, and in either case Congress must be notified. It is clear that inspectors general have created a large apparatus of compliance monitoring and bureaucratic reporting, and have used a great deal of paper; what is harder to assess is whether they have been effective at promoting executive accountability, either to Congress or to the citizenry. The leading systematic study concludes that "the Inspectors General have been more or less effective at what they do, but what they do has not been effective. That is, they do a relatively good job of compliance monitoring, but compliance monitoring alone has not been effective at increasing governmental accountability. Audits and investigations focus too much on small problems at the expense of larger system issues."
>
> (Posner and Vermeule, 2010, pp. 86–7, citing Fields, 1994, p. 505 [reviewing Light, 1993])

The notion of performance indicators of the anti-corruption effort needs to be clarified. We have had some hints and seen some interesting indicators, like the conviction-to-prosecution ratio in the case of the UK's SFO, figures on convictions in the US Office of Government Ethics prosecution report, and measures of output (total number of cases entering the institution) in Italy, the UK and others. The general notion of performance measurement needs further analysis, however. There is an inherent difficulty in measuring corruption per se, and some institutions resort to measuring their output as a proxy for effectiveness in the field. The next chapter, about measurement, is important. We also need to listen to the critics: anti-corruption plans are denounced as being merely paperwork, while the biggest challenge appears to be the widespread understanding of corruption as maladministration: this cannot be addressed without looking at the national context and at the private sector. There is a need to understand the orders of magnitude of the eco-nomic and social contexts in order to appreciate the scale of anti-corruption operations.

An opportunity exists to leverage the analogy between public and private corruption, as highlighted already in this book. From what we have learned in Part I, the anti-corruption plan emerges as an instrument with potential. However, it has its limits: it looks at intermediate variables only; it does not look at the entire spectrum of an organization's activity. Let us then widen our horizons in Part II. We will look at what surrounds the anti-corruption organization and at the effectual reality the anti-corruption organization is meant to create. We will do this by broadening the scope and looking at the context of the anti-corruption effort and at other components of anti-corruption action. This involves a closer examination of the object of anti-corruption action: the public administration complex itself and its economic and social context. We need to look at the dynamics of sustained collective action on the phenomenon of corruption, the supporters and the detractors of such efforts, the general public and the stakeholders.

Notes

1 Federalist Paper no. 10 (1787).
2 *Es gibt noch Richter in Berlin!* ("There are still judges in Berlin!") is a saying that attests to ordinary people's faith in justice (and their country's judicial system). King Frederick the Great of Prussia, so the story goes, ordered the demolition of a mill that obstructed the view from his palace, prompting these words from the poor miller. The judges ruled in favour of the miller and ordered the king to rebuild the windmill and pay compensation.

Part II

Broadening the view
Adding contextual elements

In Part I, the anti-corruption plan emerged as a potential instrument for documenting corruption and anti-corruption efforts and for promoting international dialogue. In Part II, the scope is broadened to look at the context of anti-corruption efforts and at the other components of anti-corruption action, such as its stakeholders and possible means of measurement.

The social and economic context of the anti-corruption effort

In previous chapters, public administration has been discussed in some detail but not as a uniform body. We have spoken about municipalities, regions, professional associations and other kinds of public organizations. We have spoken about ministries, independent authorities, regional executives, municipalities, local health care units and other bodies of national health care services. We have also mentioned parliament and its parliamentary questions sessions. In this chapter we will describe public administration in more detail and provide Italy's general social and economic context. We will refer to the country's absolute numbers. We will offer a description of Italian public administration and of the wider public sector that is subject to Authority action. This is what we have already referred to as the "perimeter" of the anti-corruption effort. There will also be a description of the general sphere in which the public sector operates: the country, its economy and its social fabric. By doing this we want to avoid the trap of devoting too much space to the intervention instrument (the Anti-corruption Authority's activities) and too little to the area this instrument is supposed to address (the country's public sector and its overall economy and society). In doing so, we must also remain aware that corruption is a phenomenon affected by both the internal and the external contexts of public administration itself.

In the following sections we will illustrate these concepts and provide the quantitative data that may help in painting a background picture for the anti-corruption measures described in the preceding chapters. It will begin with some general data about Italy, contextualizing the country internationally, and follow with information about Italy's social structure and regional differences. Then we present detail about the structure of Italy's public administration and finish with an overview of Italian public expenditure.

The data provided here is mostly financial; however, remaining aware that corruption is not only about money, as explained below, we also present data on people, public employees and the country's citizens in general. Besides money, corruption is also about other forms of exchange: trading of influence, time, positions and resources in general. Having said that, the monetary side of things is a reliable place to start looking for corruption.

We aim to present here an idea of the structure of Italy's society and economy. There is no intention to study trends over time, nor is this an analysis of the country's economy in general. This is a static view, a snapshot of the country in the 2010s. The structure and the quantitative value of the data is not subject to rapid change year on year. Due to the variety of data provided, the years referenced might differ: this does not appear to impair the analysis. The data is intended to provide an idea of the country's size and, as such, enable international readers to appreciate how it relates to their own reality.

Quantitative data is also provided with the awareness that, as well as relative size, absolute size matters too. Governments—cabinets and their aides— tend to be of about the same size, no matter the size of the country. Authority, energy, priorities, political impulse and leadership are limited resources, although varying among individuals. The public's attention is also limited: limited by how much information each individual can process in his or her lifetime. Implementation and execution, on the other hand, are a function of the size of the country and of government effectiveness. The complexity of phenomena related to government, such as corruption, appears to follow the rules of collective action (Olson, 1965), implying "diseconomies of scale", and collective action may not even coalesce at all due to free-rider phenomena and the like. Population size is the criterion for choosing countries to compare to Italy. The idea is to paint a picture of Italy relative to those countries it might be reasonably compared to or those countries that people in Italy and its European neighbours actually compare it to.

So how does Italy compare to its neighbouring countries? We also pose a bigger question: what countries could Italy reasonably and helpfully compare itself to? A further aspect the data is meant to describe concerns the centralized or decentralized character of public administration. In general, the data provided constitutes a proposed set of relevant indicators for the activity of preventing and containing corruption.

5.1 Italy in the international arena

Italy is a peninsula set in the middle of the Mediterranean Sea, between continental Europe and Africa. It has 60 million inhabitants in a territory of approximately $300,000\,\text{km}^2$. Italy was unified and became an independent country in 1861, as a constitutional monarchy. It has been a parliamentary republic since 1946. Italy was one of the six signatories of the 1957 Treaty of Rome which gave birth to what is now the European Union. Italy is a diverse country, rich in variety of population, geography and sectors. Likewise, its public administration is very diversified which has an impact on the definition of the context and perimeter of the Authority's actions.

It has already been mentioned that Italy has a €1.6 trillion economy. Of its 60 million inhabitants, 23 million are employed. Public administration employs 3.3 million, and 4 million work within the wider public sector which is within the perimeter of the Authority's action. The depressing effect of corruption on the economy is estimated at no less than 3% or about €60 billion per year of "foregone" GDP. Italy's population is comparable to that of France and the UK, which are both countries with which Italy compares itself. Traditionally, Italy has also always compared itself to Germany, a country with a one-third larger population. These countries appear just above Italy in the global ranking of absolute GDP (see Exhibit 5.1). Although Italy's per capita GDP is about 60% of its European counterparts, its social and governance indicators compare much less favourably. It can be observed that the government effectiveness indicator for Italy is less than one-third of that of the other countries.

One figure that is possibly a consequence of low government effectiveness is that Italy has the lowest employment percentage over total population of all the countries we are comparing it to (see Exhibit 5.2).

Note that in Exhibit 5.2 Italy's employed population is given as higher than the 23 million stated earlier; this is due to a different source being used which includes about a further one million employed that are generally not

Exhibit 5.1 Government effectiveness and GDP per capita

	Government effectiveness*	GDP per capita (current US$)**	
	2014	2014	2015
France	1.40	42,955	36,526
Italy	0.38	35,396	30,171
United Kingdom	1.62	46,783	44,305
USA	1.46	54,598	56,469
Australia	1.59	62,099	56,408

Sources: * World Bank (2014) – range is between +2.5 and –2.5; ** World Bank (2015a).

Exhibit 5.2 Employment as a percentage of total population

	Employment 2014	
	Total (millions)	% population
France	27.064	40.9
Germany	42.648	51.76
Italy	24.366	40.1
Japan	65.253	51.0
United Kingdom	30.639	47.5
USA	148.595	46.6
Australia	11.670	49.0

Source: OECD (2014).

included in the domestic statistics. This discrepancy does not alter the message, however.

The significant discrepancy in the government effectiveness indicator requires us to think again, and look for countries with similar government effectiveness performance and of a similar size, still within Europe. Attempting to match these criteria, comparison was tried between Italy and Poland, Spain and Turkey (see Exhibit 5.3).

Italy shows the same level of government effectiveness as Turkey and lags behind both Poland and Spain. To some, these results might appear counter-intuitive but this is what the data tells us. (Maybe there is a basis here for questioning indicators that are opinion-based rather than fact-based.)

Our previous comment, about the correlation of employment and government effectiveness, is validated here, as countries with lower government effectiveness have a comparable rate of employment over total population (see Exhibit 5.4).

Looking at all this data, we can observe an Italian anomaly among the nine countries. Italy appears to have a high per capita GDP compared to its very low government effectiveness. One could venture the hypothesis that the low evaluation of the Italian government effectiveness is heavily affected by Italy's large public-debt-to-GDP ratio, which is in the neighbourhood of 135% (see

Exhibit 5.3 Government effectiveness and GDP per capita

	Government effectiveness*	GDP per capita (current US$)**	
	2014 estimate	2014	2015
Italy	0.38	35,396.0	30,171.0
Poland	0.82	14,337.2	12,494.5
Spain	1.15	29,718.5	25,831.6
Turkey	0.38	10,303.9	9,130.0

Sources: * World Bank (2014); ** World Bank (2015a).

Exhibit 5.4 Employment percentage over total population

	Employment	
	2014 (millions)	% population
Italy	24.366	40.1
Poland	15.755	40.9
Spain	18.163	39.1
Turkey	26.820	35.0

Source: OECD (2014).

Exhibit 5.5). Conversely, one could say Italy's high level of per capita GDP corresponds to a form of wealth "borrowed" from future generations.

Some unpicking of Italy's low government effectiveness evaluation can be done by looking at data about doing business in Italy (see Exhibit 5.6).

This alternative source does not go into detail about low "government effectiveness"; similarly, "inefficient government bureaucracy" is not expanded on and appears to be a different wording for the same idea. It is also important to note that OECD reviews report on opinion-based indicators.

5.2 The social structure and regional differences

Italy's social structure has been comprehensively studied by scholars in the post-World War II period and it has given rise to such sociological concepts

Exhibit 5.5 Public debt as a share of GDP in selected European countries

France	97.5
Germany	71.1
Italy	135.4
Spain	0.5
Turkey	32.9
UK	87.7

Sources: Statista (2016); Turkey data from trading.economics.com.

Exhibit 5.6 Most problematic factors in doing business in Italy

	% of responses
Inefficient government bureaucracy	19.9
Tax rates	18.7
Access to financing	16.1
Restrictive labour regulations	11.1
Tax regulations	8.6
Corruption	7.2
Policy instability	5.8
Inadequate supply of infrastructure	5.5
Insufficient capacity to innovate	2.0
Crime and theft	1.7
Government instability/coups	1.0
Poor work ethic in national labour force	0.8
Inflation	0.6
Inadequately educated workforce	0.5
Poor public health	0.4
Foreign currency regulations	0.2

Source: World Economic Forum (2014, p. 222).

as "civicness" (Putnam, 1993) and "amoral familism" (Banfield, 1967). According to Putnam, "civicness" accounts for the differences between the quality of government in the north of Italy and its south. Civicness is equivalent to social capital; it is a matter of historic circumstance that the northern regions were structured around local communities, the *comuni* (commons), whereas the southern regions were still structured according to the feudal institutions of large land ownership (*latifondo*).

The social fabric of the southern regions had been studied before Putnam from a more anthropological point of view, and amoral familism had been identified as a pseudo-code of ethics whereby what happened outside the structure of the extended family was beyond the concerns of ordinary ethical behaviour, which in turn was strictly observed within the boundaries of the family.

It should be noted that, since it was first studied in the late nineteenth century, the country appears to show a geographic pattern in most of its indicators where the south performs worse than the north. Crime in the south is only reported 50% of the time. As a corollary, and just a single example, fraud on auto insurance claims is eight times as high. Organized crime also appears to be more vigorous in the south (Ricolfi, 2007, p. 98 and p. 121). Exhibit 5.7 shows the distribution of corruption-based crime across the country's 20 regions. The regional differences mentioned in this chapter's introduction are clearly evident.

Exhibit 5.7 Average corruption cases in the Italian regions (average between years 2004 and 2010, per 100,000 inhabitants)

Valle d'Aosta	12
Piedmont	5
Lombardy	3
Veneto	5
Trentino-Alto Adige	5
Friuli-Venezia Giulia	5
Emilia-Romagna	2
Tuscany	5
Umbria	8
Lazio	5
Marche	5
Abruzzo	8
Molise	12
Campania	7
Basilicata	14
Apulia	8
Calabria	20
Sicily	10
Sardinia	5

Source: elaboration on Fiorino and Galli (2013, p. 46). The regions are listed from the north-west through the north-east, to the centre, south and islands.

The data presented above is intended to convey the background of Italy and its anti-corruption effort to the international reader. As such, we have introduced Putnam and Banfield, recognized international scholars, who have made Italy the subject of their research. Our book is not a study about the causes of corruption, however.

5.3 The structure of public administration

We will now describe the structure of Italian public administration and the wider public sector in order to understand the Authority's orbit of supervision. A description of the organizational units is followed by data about the people who work in them, and then finally an illustration of the funds involved and their significance within Italian GDP.

There are three levels to Italian public administration's organizational structure: central government, the 20 regions and the local authorities (which include 110 provinces and metropolitan cities and over 8,000 municipalities). Complications to the three-level model can be created by including some intermediate entities but those are disregarded here. It is important to mention, however, that Italy's own legislation—and to some extent Italy's public administration—is affected by legislation, regulation and financial support from the supranational EU level.

The organizational units of public administration number over 10,000. As mentioned, the large majority of these (over 8,000) are municipalities, which may be quite fragmented and include only a few hundred inhabitants. Under this reckoning, large organizations—e.g. central government ministries—count as one; regional administrations number 20; regional health care units about 400. Relationships between the diverse levels of government are shaped via a model of uncertain decentralization or federalism. The system occupies a place between the classical models of centralized France and decentralized Germany. Further insight into this system will be offered in the following text, where we illustrate the different levels of public expenditure pertaining to central government with respect to the regions and local communities.

Alongside the organizational units of public administration, the public sector includes 7,684 private organizations under governmental control. Of these, approximately 4,400 are corporations or SOEs. Among these corporations we may recognize names known to the international investor or reader: Eni, the oil concern; Enel, the utility company; Poste Italiane, the postal service corporation; Trenitalia, the railway transportation concern; and A2A, Hera and Acea, the utilities owned by the municipalities of large cities like Milan, Bologna and Rome. The SOEs represent an opportunity for outsourcing activities of public interest: core public functions, public utilities and private services provided in the administration's interest.

However, it appears that SOEs listed on the stock market are—counterintuitively—not subject to anti-corruption regulation, with only the unlisted ones subject to Authority supervision. Listed SOEs include: ANAS, road-works and maintenance; Invitalia, management and economic services, under the supervision of the Ministry for Economic Development; INAIL, work-place safety, under the supervision of the Ministry of Labour; and INPS, social security administration, also under the supervision of the Ministry of Labour.

For the purpose of anti-corruption activities, data about the number of organizational units should be augmented with data about the number of ten-dering units for the purchase of public works, and the number of procure-ment units for the purchase of goods and services, that are located within these organizations. The tendering units number about 37,000, more than the sum total of organizations since many organizations have more than one ten-dering or procurement unit. Legislation appears to be in the offing to cut this number drastically to no more than 1,000. Procurement units have been reduced to 35 whereas work on tenders is still to be done (Donato, 2016).

Having looked at the organizational structure of the Italian public adminis-tration and of the wider public sector of which it is part, we will now intro-duce the people who work there. Exhibit 5.8 shows the number of people working in Italian public administration by sector.

The total number of employees in public administration is over 3.3 million, over 3 million of which are tenured. To this number we must add those

Exhibit 5.8 Employees in public administration, 2013

Schools	1,028,231
Music schools	9,441
Central ministries	163,369
Presidency of the Council of Ministries	2,359
Tax agencies (internal revenue service)	52,529
Fire departments	35,038
Police corps	316,717
Armed forces	185,325
Judiciary	10,425
Diplomacy	910
Prefect track	1,277
Correctional facilities managers	356
Non-economic public organizations	48,983
Research organizations	24,160
Universities	107,352
Health care service	702,475
Regions and local organizations	527,168
Regions with special statute	105,713
Independent authorities	2,287
Other organizations (art. 60 and art. 70)	12,384
Total	**3,336,499**

Source: Ministry of the Economy, www.contoannuale.tesoro.it.

employed in the wider public sector: state or municipally owned corporations and other entities under public control. Conservatively, their number can be set at about 700,000. Therefore, the wider public sector in Italy makes up about 4 million people out of a total employed population of 23 million.

The public administration total in Italy appears to be lower than in corresponding countries (see Exhibit 5.9).

Personnel is a key factor in the Anti-corruption Authority's view of public administration. Within this perspective, the Authority's President—speaking during a meeting with Corruption Prevention Officers on 24 May 2016—quoted the May 2016 IMF document on corruption:

> While designing and implementing an anti-corruption strategy requires change on many different levels, the Fund's own experience in assisting member countries suggests that several elements need to be given priority. These include transparency, rule of law, and economic reform policies designed to eliminate excessive regulation. Perhaps most importantly, however, addressing corruption requires effective institutions. While building institutions is a complex and time consuming exercise that involves a number of intangible elements that may seem beyond the reach of government policy, the objective is clear: the development of a competent civil service that takes pride in being independent of both private influence and public interference.
>
> (IMF, 2016)

In contextualizing public administration within the international arena, we should also look at remuneration of public employees. International organizations often underline the importance of public employees' "adequate remuneration" as a safeguard against corruption. The UNCAC states:

> Article 7. Public sector—1. Each State Party shall [...] endeavour to adopt, maintain and strengthen systems for the [...] civil servants [...] *(c)* That promote adequate remuneration and equitable pay scales, taking into account the level of economic development of the State Party.

Exhibit 5.9 **Employees of public administration in selected European countries (2013, millions)**

France	5.600
Germany	4.635
Italy	3.336
UK	5.319
Spain	2.937

Source: La spesa per redditi da lavoro dipendente: confronto tra Germania, Francia, Italia, Regno Unito e Spagna.

It can be asserted that Italy is fully compliant with this requirement: public salaries have consistently averaged 27% above private salaries since the 1980s according to a report by IMF manager Carlo Cottarelli, who at the time was the Italian spending review czar (Cottarelli, 2014, p. 16).

This relative higher level of public administration salaries does not appear to be much of an exception globally, however, as shown in Exhibit 5.10 using international data from the IMF on the same subject.

Describing the trend in Europe, the European Central Bank (2016) stated: "Overall aggregate data show that the euro area government wage differential with respect to the private sector increased from 20% in 2007 to 25% in 2009 and subsequently fell to 23% in 2014 owing to fiscal consolidation measures."

Cottarelli also revealed the levels to which Italian managers' remuneration exceeds that of top managers elsewhere in Europe (see Exhibit 5.11).

The same report shows analogous data for other levels of management. Exhibit 5.12 shows how public managers' remuneration, taken as a whole, relates to their countries' per capita GDP.

A commendable aspect of the Cottarelli methodology is that it compares cross-national salaries based on absolute values rather then on variations over time. Cottarrelli also compares each country's figures with their own per capita GDP. Such a methodology is far from standard in the literature, where salaries of public employees are frequently compared only according to their increments or in absolute values, irrespective of each country's wealth and income (Franzini, Granaglia and Raitano, 2016).

The data presented in these Exhibits applies to public administration, i.e. directly employed by government. It can be safely inferred that figures for the wider public sector are no less favourable than they are in the core public administration.

5.4 Public expenditure in detail

Having examined public administration organizational structure and personnel, we will now turn our attention to the funds available to these organizational units. Public expenditure is in fact part of the quantitative context of the specific measures taken by the Anti-corruption Authority, sector by sector. Exhibit 5.13 shows public expenditure as a share of GDP in the comparison countries. Italy is shown to have somewhat higher public expenditure.

With gross domestic product (GDP) in 2013 a little over €1.6 trillion and public expenditure about 50% of GDP (Exhibit 5.13), we obtain a public expenditure total of about €800 billion, as corroborated in the bottom line of Exhibit 5.14. Note that the country's GDP in the 2010s stayed basically the same for several years. It is interesting, then, to see how this €800 billion is spent in absolute terms.

Exhibit 5.10 Global public-sector wages relative to other economic sectors

	Number of countries	Ratio of avg. PA wage to per capita GDP	Ratio of PA to financial sector	Ratio of PA to manufacturing sector
Africa	3	1.3	0.7	1.8
Asia and Pacific	7	1.4	0.9	1.4
Europe	28	1.4	0.7	1.3
Western hemisphere	11	1.4	0.8	1.3
Middle East and Central Asia	8	1.2	0.5	1.3
European Union	17	1.3	0.7	1.3
Low-income countries	4	1.9	0.7	1.4
Middle-income countries	35	1.4	0.6	1.4
High-income countries	18	1.2	0.8	1.3

Source: IMF (2010).

Note
PA = public administration.

Exhibit 5.11 Amounts to which Italian top public managers' remuneration exceeds those of other countries

UK	48.9%
Germany	154.1%
France	96.1%

Source: Cottarelli (2014).

Exhibit 5.12 Top public managers' remuneration expressed as multiples of their own country's GDP per capita

UK	8.48
Italy	12.63
Germany	4.97
France	6.44

Source: Cottarelli (2014).

Exhibit 5.13 Public expenditure as a share of GDP in selected European countries

France	57.1
Germany	44.3
Italy	50.5
Spain	44.8
UK	45.5

Source: Tuzi (2016, p. 108).

Exhibit 5.14 Public administration expenditure categories, 2010

	€ (million)	%
Public consumption	328.607	41.5
Pensions	240.000	30.2
Welfare payments	69.947	8.8
Industrial policy	15.330	1.9
Other current	15.579	2.0
Interest	70.152	8.8
Capital expenditure	53.899	6.8
Total	**793.514**	**100.0**

Source: Giarda (2012, p. 9).

It is worth noting that some of the sources of this national economic overview are those that were commissioned by spending review czars: Professor Piero Giarda was the first, preceding Carlo Cottarelli. In 2014, the spending review was entrusted to the Renzi Cabinet, under Yoram Gutgeld.

Exhibit 5.15 Breakdown of public expenditure by function, 2012 (as a percentage of GDP)

General services	9.1
Defence	1.4
Law enforcement	1.9
Economic affairs	3.4
Environmental protection	0.9
Housing and land management	0.7
Health care	7.3
Culture	0.7
Education	4.2
Welfare and pensions	21.0
Total	**50.6% of GDP**

Source: Tuzi (2016, p. 113).

Turning to capital expenditure, the last row of Exhibit 5.14, it breaks down into construction work (buildings), which is twice that of roads, which in turn is equal to "miscellaneous work" (i.e. 50%, 25% and 25% respectively). Cottarelli's spending review also revealed investment in Italy to the tune of 3% of GDP, which is consistent with 6.8% of capital expenditure in Exhibit 5.14. It is interesting to note that the equivalent amount in Germany is 2%. Cottarelli hinted that Italy's high figure might be a consequence of corruption (Cottarelli, 2016).

More specifics of the country's public expenditure are revealed in Exhibit 5.15. Note that health care, welfare payments and pensions receive a larger share than in corresponding countries. They appear to be over-represented when viewed through the "Washington consensus" lens.

5.5 The Anti-corruption Authority's framework

Within the general national framework outlined in previous sections, the Anti-corruption Authority sets out its own parameters for activity, calculating its own "relevant market". The Authority's 2015 Annual Report shows a total value of contracts (exceeding €40,000) of €117.3 billion (see Exhibit 5.16).

A specific aspect of our methodology is to contextualize the "micro" activity of an organization within its general environment, and similarly to compare documents produced by an organization to other sources. In doing this we can affirm that the Authority appears to have a clear understanding of its environment, as evidenced by the framing of its activity within the relevant market. The numbers compare: the Authority says a total "market" of €117.3 billion (Exhibit 5.16). The Giarda figure (Exhibit 5.14) for public consumption is €328.607 billion which, added to capital expenditure (€53.899 billion), takes us to €382 billion. From this, about €200 billion should be deducted for public administration employee remuneration. The

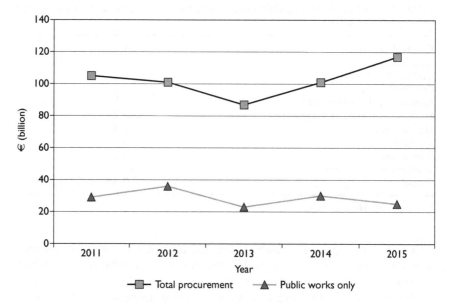

Exhibit 5.16 Value of contracts (exceeding €40,000).

Source: ANAC (2016, 14 July), p. 157.

Authority's relevant market is therefore located in the €182 billion of public capital plus consumption expenditure, net of salaries.

A final point relates to the figure for local public administration, which is 30% of the total. This is somewhat misleading because, as we have seen, total expenditure includes welfare payments, pensions and interest on the sovereign debt, which are marginal to the concerns of this book. When we consider only funds the disbursal of which is discretionary on a year-on-year basis, then we uncover an interesting result: central government accounts for 35.6% of expenditure while local government accounts for 62.4%.

The data about the "relevant market" shows that the Authority brings a systems view to the phenomenon of corruption. A systems view implies that the Authority trains its spotlight on the whole economy and on all potential public administration corruption within its public administration and wider public-sector perimeter. A systems view also implies the possibility of changing the rules, which in fact is what the Authority does via recommendations and interaction with the cabinet and parliament. Upholding the legitimacy of the rules does not imply that the rules cannot be made more effective.

To pick a particular example: in the 2016 Anti-corruption Plan, the Authority concentrated on, among other areas, health care. The previous Exhibits are useful here because they show the relevance of the public health care system in the Italian economy: €110 billion in 2012 or 7.3% of GDP

(Exhibit 5.15). We also have sector-specific numbers that show us that the total number of government health care employees was 702,475, including doctors, nurses, aides and administrative personnel. They represent more than 20% of the total number of government employees and about 3% of total employment.

These numbers provide a basic minimum quantitative framework of the input resources that government is providing to this sector. We should make it quite clear that input resources are the most basic and crudest indicator of potential corruption. In reality, corruption—and maladministration—occurs in the development of service processes and has an impact on the outcome of such services. The outcome has a quantitative measure way higher than the input resources, as will be seen in Chapter 6 ("Measuring the impact of the anti-corruption effort").

The Authority's focus on expenditure related to health care and procurement of public works reveals not only a priority approach (i.e. objective) but also a subjective approach: the National Agency for Health Care (Agenas) was more collaborative than other public administrations. Nor should we assume that the Authority focuses only on public expenditure and direct governmental outlays, or "cash out" operations. It is worth noting that government expenditure (procurement) is only one area of potential corruption. Corruption may in fact occur where there is no direct government expenditure; it may, for instance, take the form of "speed money" or "evasion of work", which are elements of "maladministration". Within public expenditure, a focus on health care and public procurement may reveal an optimizing strategy as those sectors comprise a large share of public expenditure.

5.6 Conclusion

The numbers presented in this chapter reveal a level of complexity in Italy: of structure, organization and expenditure. The hypothesis is that the basic factor driving this complexity is the country's size of population relative to its level of social and economic development. This is why, at the beginning of this chapter, we selected for comparison countries with similar populations and per capita GDP. Comparisons with countries of different sizes would not deliver insights for future management, we contend.

This chapter has aimed to highlight the diversity of Italy's external and internal public administration. It is hardly unusual from a global point of view. However, such a variegated landscape of organizations and institutions corroborates the Authority's premise that penal deterrence is not equal to the task of containing corruption. The threat of prosecution and jail will not be enough to reduce corruption to the marginal phenomenon that it is seen to be in the countries to which Italy compares itself. On the other hand, the Authority is not going to use its power to impose a new homogenization within the country's public administration and within the country's economy.

The Authority is concerned only with practical results; there is no hidden agenda.

Examining Italy's public administration in a global context, on the one hand it can be placed at the margins of the classical Western countries, much closer to France and Germany than to the UK and USA; on the other hand, we must take into account the low performance of Italy's public administration and its government effectiveness, which are the Authority's very *raison d'être*. Therefore, we are led to concede that Italy can also be placed at the margins of Riggs's "polyarchal competitive model" for developing countries (Heady, 2001, p. 372). Riggs describes this model as the one in which developing countries come closest to Western models. The model evinces the prevalence of bureaucracy in the political system, which in Italy we can observe through the juridical status of the bureaucracy as distinct from the political system. We can also recognize in Italy many of the characteristics authors have observed in developing countries' public administration: "unequal distribution of services, institutionalized corruption, inefficiency in rule application, nepotism in recruitment, and a pronounced gap between formal expectations and actual behavior" (Heady, 2001, p. 29). We concur with LaPalombara in appreciating "the difficulty in restricting bureaucracy to an instrumental role" (Heady, 2001, p. 100).

It is to be hoped that this chapter has contributed to developing a top-down risk map of the Italian public sector—enumerating measurements of weightings and differences among different public administration organizations—which might assist in improving the efficiency of the corruption prevention process. The figures validate the priorities the Authority has set for itself, namely public expenditure in health care and public works.

Having investigated the structure and some elements of society and the economy, the tough task ahead of us is to devise an estimate of the corruption actually affecting the country and its public sector. This is the subject of the next chapter.

Chapter 6

Measuring the impact of the anti-corruption effort

6.1 Introduction

The previous chapter presented the context of the Italian economy, society and public administration. This chapter aims to provide an orientation in that landscape. It elaborates on the metrics of corruption and of the anti-corruption effort, so that the Anti-corruption Authority's activity can be assessed in the context of Italy's reality. We will begin by reviewing the few available current measures of Italian corruption in the country: by the OECD, the Italian Court of Auditors, the EU, the European Council's GRECO (Group of Countries against Corruption) and Italian institutional sources. We will also present academic sources for corruption measurement. After reviewing possible metrics for final impact, this chapter will further consider measurement by looking at intermediate variables. Finally, it will provide a taxonomy of corruption as a potential tool for further investigation into the domain of risk mapping and to enable a specification of what constitutes maladministration.

6.2 Foregone GDP

Let us begin by citing the OECD Integrity Review of Italy:

> The level of perceived corruption in Italy has risen continuously since 2007, while trust in the government's ability to control corruption has declined steadily since 2000. Italy's national audit office, the Court of Auditors estimates that the cost of corruption in Italy in 2011 was approximately EUR 60 billion, the equivalent of the federal government deficit in the same year.
>
> (OECD, 2013)

The citation from the Integrity Review of Italy lays the ground for a first measurement taxonomy as it mentions both the "cost of corruption" and public "perception". First of all, we need to define what is meant by the cost of corruption, and it soon becomes apparent that "cost of corruption" measurements are actually just a top-down monetary quantification of public perception. This is a general weakness of comparative public administration: available metrics about the quality of government are usually framed as measures of expenditure or measures of opinion and no mention is ever made of measures of qualitative output or outcome. This is one sign of "comparative public administration not being a well worked field" (Ackerman, 2000, p. 706). Measures of efficiency and effectiveness are a staple of public policy research and of selected sectoral statistics, but when it comes to comparative country studies and specific fields—such as corruption—synthetic indices (no matter how imperfect and improvable) are derived only from public opinion rather than hard data. This makes it almost impossible for countries to identify what to do to improve their management and policies, short of embarking on a public relations and communication programme to change opinion without changing effectual reality. Nonetheless, the figures are powerful and tell us that corruption in Italy is pervasive and cannot be ignored. Something must be done.

Let us analyse what is meant by the cost of corruption by examining statements from different actors quoting and commenting on these figures. Note that the OECD Integrity Review quotes an Italian Court of Auditors "estimate", by President Luigi Giampaolino. However, press coverage of a 2012 event at which this figure was announced tells a different story (*Panorama*, 2012). On 15 February 2012, at the inaugural ceremony of the judicial year at the Court of Auditors, a public prosecutor, Maria Teresa Arganelli, also said that in 2011 the sum total of first degree judgement was only €75 million. Following appeals, that number dropped to €15 million.

It seems a misunderstanding about definitions might have been taking place. On the one hand, we probably have a measure of "foregone GDP" and on the other hand we have a reading of that same figure as the "monetized quantity of annual corruption in Italy". And there is yet a third measure: the "sum total of first degree rulings".

At this point we now have an insight into what can be meant by "cost of corruption". By "cost" of corruption, economists mean the loss of GDP due to corruption, i.e. the non-growth, the economic development that was prevented from happening because of corruption. This is the ultimate measure of the impact of corruption. Conceptually, this is an opportunity cost (*lucro cessante*): missed development. This is what literature calls "foregone GDP".

At this point three different types of metrics have been introduced: foregone GDP, monetized quantity of annual corruption or "amount of bribes paid" (see the European Parliament's analysis later in this chapter), and the sum total of first degree judgement. There is actually one further measure to consider: the "sum total of public expenditure affected by bribes". There is no estimate available to put a number on this value. However, this is the most interesting metric for anti-corruption activity because this figure can actually be compared with the sum total of public expenditure given in

Chapter 5, in which it was stated that "the Anti-corruption Authority sets out its own parameters for activity, calculating its own 'relevant market'" and the "Authority's 2015 Annual Report shows a total value of contracts (exceeding €40,000) of €117.3 billion" (see also Exhibit 5.16). This is the crux of the orientation function we introduced at the beginning of this chapter: to be able to find the locus of pervasive corruption in the country, and specifically in public expenditure, seeing as the Authority's purpose is to prevent corruption in public administration. We should probably also reiterate here that corruption also takes place in the private sector, but that is not the within the Authority's scope.

Summarizing the four types of metrics introduced in this section, they appear to intervene at quite different levels in the phenomenon of corruption. Institutions do not tell us which of these metrics are more meaningful, but nonetheless it is possible to identify a hierarchy, from the widest-ranging to the narrowest:

1   Foregone GDP: €60 billion out of a GDP of €1,600 billion;
2   Sum total of public expenditure affected by bribes: not quantified, out of a relevant market of €117.3 billion;
3   Monetized value of annual corruption or "amount of bribes paid" (see the European Parliament's analysis that follows): not quantified;
4   Sum total of first degree rulings: €75 million.

A 2016 RAND study under the aegis of the European Parliamentary Research Service (EPRS) helps us hone in on the third metric: the monetized quantity of annual corruption or "amount of bribes paid". The study has an estimate for Italy of US$105 billion of "average annual reduction in GDP" (European Parliament, 2016, p. 44). The study then explains the difference between the "average annual reduction in GDP" (or "foregone GDP") and the amount of bribes paid:

> the estimate [of foregone GDP] is substantially higher than the global value of bribes paid estimate circulated by the World Bank [see following]. However, it is important to note that our estimates, by measuring the foregone annual GDP, include at least largely the indirect effects that accrue from high levels of corruption and do not focus solely on specific cost elements such as bribes paid.
>
> (European Parliament, 2016)

The World Bank brief—cited by the EPRS RAND study—quantifies the amount of bribes paid globally:

> Businesses and individuals pay an estimated $1.5 trillion in bribes each year. This is about 2% of global GDP—and 10 times the value of overseas development assistance. The harm that corruption causes to development is, in

fact, a multiple of the estimated volume, given the negative impact of corruption on the poor and on economic growth.

(World Bank, 2017)

According to the Trading Economics website: "The Gross Domestic Product (GDP) in Italy was worth 1849.97 billion US dollars in 2016. The GDP value of Italy represents 2.98 percent of the world economy."[1] So, if Italy represents about 3% of global GDP, its 3% share of bribes paid would be equivalent to €45 billion, which appears to be too much in relation to the amount of foregone GDP (estimated at between €60 billion and US$105 billion). The World Bank estimate of bribes paid appears to be excessive. Last but not least, it must be noted that the RAND study cited previously is a quantification of global perception indices.

These estimates should not be seen as fanciful, however, because a 1%-of-GDP estimate of foregone GDP has been confirmed by scientific research. Galli estimated 0.8% (2013), and Golden and Picci (2006) have a figure between 0.5 and 1.0% of GDP.

Referring to a more detailed methodology, albeit dated, a study by the Italian Industrialists Association gives estimates in the same order of magnitude regarding the impact of public administration on the Italian economy:

The hostile relationship between public administration and business firms, spanning from slow justice to delays in payment, is a decisive obstacle to the deployment and growth of any economic activity. The Italian GDP could grow by as much as 30% by the year 2030, as shown in the following exhibit.

(Italian Industrialists Association [Confindustria], quoted in Lapiccirella, 2015, p. 105)

Shown here as Exhibit 6.1, the referred-to exhibit is on a different wavelength but it points to the same reality and the same phenomena. It does not speak of corruption explicitly but it is linked very closely to it.

Exhibit 6.1 **Confindustria estimates regarding the impact of public administration on the Italian economy**

---

1 Lower administrative burden on business: % of GDP +4.0 (or +62.9 billion euro)
2 Better infrastructure: % of GDP +2.0 (or +31.4 billion euro)
3 Better human capital: % of GDP +13.0 (or +204.4 billion euro)
4 Competition: % of GDP +11.0 (or +172.9 billion euro)
5 Total: % of GDP +30.0 (or +471.7 billion euro)

Source: Lapiccirella (2015, p. 105).

Note
Percentages of GDP are based here on 2008 GDP.

On measures of perception, the general 2014 European Commission report on corruption does not include objective indicators, and only one table on perception of countries' corruption.

> According to the 2013 Special Eurobarometer survey on corruption, (45%) of the Europeans interviewed believe that bribery and the abuse of positions of power for personal gain are widespread among officials awarding public tenders. The countries where respondents are most likely to think that there is widespread corruption among officials awarding public tenders include the Czech Republic (69%), the Netherlands (64%), Greece (55%), Slovenia (60%), Croatia (58%) and Italy (55%). Countries with the most consistent positive perceptions of officials in this area include Denmark (22%), along with Finland (31%), Ireland (32%), Luxembourg (32%) and the UK (33%).
>
> (European Commission, 2014)

The Economist said:

> Europe is riddled with corruption or so its citizens believe. Three-quarters of respondents to a Eurobarometer survey in 2013 said they felt that corruption was widespread in their country. Perceptions of corruption seem to be more sensitive to claims than facts.
>
> (The Economist, 2016b)

Likewise, in 2015, 90% of Italian interviewees said Italy is a corrupt country, whereas fewer people said they had been involved in a corruption case (Carloni, 2017).

It is worth noting that nowhere in its documents does the Authority refer to opinion-based measures of corruption.

6.3 Analysis of legislation

Resuming our review of attention to corruption in Italy from international institutions, we turn to the international evaluations of GRECO, which are mostly about the many intermediate steps taken and institutional arrangements formulated for tackling corruption, e.g. whether an anti-corruption agency was established, whether a law was passed making false declarations in financial statements a penal offence.

GRECO evaluations do include some judicial statistics related to corruption (e.g. number of cases brought to court). However, the Authority was not convinced of those statistics' usefulness in helping to measure its own effectiveness. We should be aware of the use of numbers here: in spring 2015, the public was under the impression there had been a surge of corruption cases coming from the judicial branch, leading to the question of whether those cases were an effect of a surge in corruption or merely an effect of a surge of *attention* on corruption.

Cases appearing in the news today have their origins in judicial branch activity begun at least two years earlier, such was the minimum duration of a judicial branch investigation before it surfaced in the media. So the answer to the above question could be: neither. Nonetheless, corruption had been on the government's radar for at least two years, it was a hot topic among the public and in the media, and the prosecutors may have felt appreciated in their anti-corruption efforts. An expected surge in "signalling", for instance, happened in spring 2015 when the Authority passed a resolution on whistle-blowing—an indispensable anti-corruption instrument according to the Authority.

A comparative analysis of GRECO evaluations about the UK, Germany, France and Italy did not lead to an identification of better indicators in certain countries. In other words, the fact that corruption in northern and central European countries is lower than in Italy is mostly likely true; however, it is also true that this a matter of opinion rather than a scientific fact.

Finally, the Italian National Institute of Statistics (ISTAT) appears to have been thinking about specific corruption metrics, but the only relevant statistics were produced in the domain of the judicial branch. These were used in the GRECO evaluations as indicators of corruption containment activities. Nor did the Authority come up with an explicit metric, but they did launch several initiatives to gain an understanding via research and academic support.

6.4 Academic sources

Turning our attention to academic sources, we examined the literature about corruption in Italy for objective measures of corruption that the Authority could use in helping it to orient its operations. Reports on regional differences based on judicial data (Italy is geographically and politically divided into 20 regions) are given in Fiorino and Galli (2013, Chapter 5). This appears to be the most helpful in terms of Authority operations.

Other studies may be useful in terms of discussion about negative correlations between corruption and long-term economic development (Fiorino and Galli, 2013; see also Fiorino, Galli and Petrarca, 2012); however, these are not helpful for short- or medium-term evaluation of Authority action. In general, corruption and development studies demonstrate the link between corruption and lower growth/lower development but, again, these studies are not helpful in the short and medium run as they do not measure specific, local and disaggregated variables that can vary—or can be made to vary—year on year or within a three-to-five-year period. However, such studies do corroborate the crucial point that corruption is negatively correlated with development. This is the key long-term indicator that tells us that corruption must be repressed and prevented. Klitgaard (2014) becomes increasingly more emphatic about corruption's negative correlation with development, following decades of his own sustained effort to contain corruption, which implies the assumption must not be taken for granted. In fact, it is quite often heard that corruption is a "lubricant" of business.

Relating specifically to Italy, an anecdote from the mid-1990s concerns a relatively large-scale judicial action on the corruption front which became known as the "clean hands" (*mani pulite*) case. After a series of arrests was carried out on corruption and extortion charges in public procurement, a minister of a state-owned enterprise was heard to say: "when kickbacks were there, business thrived; when kickbacks were being repressed, business was stagnant". This is not necessarily an entirely cynical statement, as we should not take for granted that an absence of corruption would be filled with short- and medium-term productive and proactive government action. In fact, government officials often avoid the temptation of corruption by resorting to inaction. In truth, even though corruption might seldom be held to account, in public administration inaction is hardly ever punished.

Meanwhile, the perception of corruption as a "lubricant" is an insider perspective—of those doing business within the "corrupt equilibrium" (Klitgaard, 1988). The lubricant argument ignores those who are left out of the system: the unknown stakeholders, the latent and forgotten groups who would have generated a more efficient economy had they been included in the system of trade with governments. This goal is obtained not only through an absence of corruption, however, but also through effective action.

Putnam (2016) argues that economic development is exclusive rather than inclusive, and we need to admit that corruption may reveal organizational issues and probably legislative ones too, as corruption is only to a limited extent a matter of personal ethics. Corruption is also an organizational product. It may reveal imperfect market conditions and under-developed market institutions and organization (Schelling, 1978). To truly eliminate corruption may imply new laws and new organization.

6.5 Intermediate variables

For all that can be said about corruption and development studies, an anti-corruption organization is still left with the problem of measuring its own short- and medium-term impact. It is likely they must resort to measuring intermediate variables that are presumably (negatively) correlated with corruption. The Authority needs short-term measurements—"high-frequency" ones: something that can vary over a single year or a trend that can be observed within a maximum three-to-four-year time-frame, bearing in mind the Authority has a six-year mandate (2014–2020). Potential shorter-term indicators need to be intermediate variables related to public administration processes. For instance, we could measure the number of tenders won by a foreign company, or the number of companies competing for a tender where multiple tenders are won by a sole competitor. Other intermediate variables might be the number (or monetary value) of procurement showing cost over-runs. Other measures could be the cost of specific items purchased by the government, e.g. one kilometre of subway. Total contracts (number and volume) is also a key variable as it is important that repression of corruption does not retard economic and governmental activities. Maybe cooperation with international authorities and foreign public administrations could be established to ascertain, for example, the presence of foreign contractors as a result of public procurement in different countries. Use might also be made of the "legality rating" of companies (*rating di legalità*) whose management was entrusted to the Italian Competition Authority (AGCM).

A further, more specific, measurement of the impact of anti-corruption action—for the sole area of procurement—is a reduction in the variance of the cost of what is purchased (Golden and Picci, 2006).

Working on intermediate variables can also be seen as a bottom-up approach to capturing and measuring corruption. Another potential bottom-up approach is drawing up a taxonomy of corruption's methods and pro-cesses. To this end, we will now review the different types of corruption, as a preliminary pointer to potential specific metrics.

6.6 Types of corruption

Corruption does not always—and maybe only even in a minority of cases—take the form of a direct exchange of money. It may often involve an indirect exchange of money or it may not even involve money at all. A taxonomy of corruption phenomena has frequently appeared in the literature, and we offer the following list here, based on Rose-Ackerman and Palifka (2016, p. 8) and Klitgaard (1988, pp. 192–3).

6.6.1 Bribery

This is what is usually understood as corruption: the

> explicit exchange of money, gifts in kind, or favors for rule breaking or as a payment for benefits that should legally be costless or be allocated on terms other than willingness to pay. Includes both bribery of public offi-cials and commercial bribery of private firm agents.

6.6.2 Extortion (or speed money)

This is known as *concussione* in Italian: "demand of a bribe or favor by an offi-cial as a sine qua non for doing his or her duty or for breaking a rule". Extor-tion is "a form of bribery where the bribe taker plays an active role. (Sometimes the rule is created by extortionist in order to exact the bribe.)"

6.6.3 Exchange of favours

"The exchange of one broken rule for another."

6.6.4 Nepotism

"Hiring a family member or one with close social ties, rather than a more qualified but unrelated applicant."

6.6.5 Cronyism

"Preferring members of one's group—racial/ethnic, religious, political, or social—over members of other groups in job-related decisions."

6.6.6 Judicial fraud

A decision based on any of the preceding types of corruption, or threats to the judge, rather than the merits of the case.

6.6.7 Accounting fraud

Intentional deception regarding sales or profits (usually in order to boost stock prices).

6.6.8 Electoral fraud

Manipulation of election results, through vote buying or threats to the electorate, or by falsification or destruction of votes.

6.6.9 Public service fraud

> Any activity that undermines the legal requirements of public service delivery even if no bribes are paid. For example, teachers might provide students with the correct answers or change students' responses on standardized tests (usually in order to ensure funding). Health care providers might prescribe unnecessary tests or invent patients to increase reimbursements. Civil servants might neglect their jobs for private sector work, steal supplies for resale, or simply not show up for work.

A further example is *fantasmas* or late reporting to work. It is interesting that the literature pays this phenomenon less attention than the others. Parsing the literature on public administration behaviour, the *fantasmas* phenomenon is part of what could be called "evasion of work", i.e. work not subject to continuous comparative evaluation, within a pluralistic context. In a complex society, work can in fact also be negative, i.e. draining resources from society and giving nothing in return. Alternatively, we could see corruption as an extreme form of bureaucratic behaviour, of which it is a core subset. The evasion of work is hence a consequence of bureaucratic behaviour: work without relevance to—or with an adverse effect on—the organization's substantive mission (D'Anselmi, Chymis and Di Bitetto, 2017).

6.6.10 Embezzlement or theft

"Theft from employer (firm, government or NGO) by the employee."

6.6.11 Kleptocracy

"An autocratic state that is managed to maximize the personal wealth of the top leaders."

6.6.12 Influence peddling

Using one's power of decision in government to extract bribes or favours from interested parties.

6.6.13 Conflict of interest

Having a personal stake in the effects of the policies one decides.

6.6.14 Tax evasion and tax arreglos (assessment)

Corrupt taxpayers (with or without help from public administration officials) do not pay taxes.

The above taxonomy is helpful in spelling out the full range of what is meant by "maladministration".

When confronted with such a variety of ways of being corrupt, it becomes clearer why the only comprehensive indicator of the impact of corruption is foregone GDP, as outlined above.

Although bottom-up approaches may lack the advantage of seeing the whole economy and capturing the whole phenomenon, they can provide insights for further study, risk assessment and specific anti-corruption measures. Klitgaard (1988), in fact, includes in his table financial estimates of the transactions that take place for each type of corruption: who benefits, who suffers, and the possible causes and cures for each type. Although bottom-up approaches may only measure inputs, and furthermore may be organization-specific, they do help create a vivid image of where the problem lies. It appears to be an effective and timely way to run a risk map, with the proviso that input measures cannot be assumed to be proxies of outcome measures. A subsequent analysis can lead to the identification of potential metrics of prevention and repression.

We will leave the final words in this chapter to Klitgaard:

> Consider two analytical points. Metaphorically, corruption follows a formula:
>
> $$C = M + D - A$$
>
> Corruption equals monopoly plus discretion minus accountability. Whether the activity is public, private, or non-profit, whether you are in Washington or Ouagadougou, you will tend to find corruption when someone has monopoly power over a good or service, has the discretion to decide whether you receive it and how much you get, and is not accountable. Second, corruption is a crime of calculation, not passion. True, there are saints who resist all temptations, and honest officials who resist most. But when the size of the bribe is large, the chance of being caught small, and the penalty if caught meager, many officials will succumb.
>
> (Klitgaard, 1997)

> Givers and takers of bribes respond to economic incentives and punishments. To reduce corruption, reduce monopoly and enhance competition. Limit official discretion and clarify the rules of the game. Enhance accountability about processes and especially about results.
>
> (Klitgaard, 2014)

In the next chapter we will analyse how the public is affected by or affects the Authority's mission, and public responses to its actions.

Note

1 https://tradingeconomics.com/italy/gdp, accessed 20 October 2017.

Chapter 7

Origin and support of the anti-corruption effort

The stakeholders

In this chapter we will look at how the Italian Anti-corruption Authority is received by specific groups in Italy and abroad. We are going to refer to such groups as "stakeholders", which is the current term used in management to identify those who are in any way affected by or have any kind of interest in the Authority's activities. The subject of stakeholder management and communication is closely linked to the subject of the impact, and evaluation thereof, of Authority activities and its effectiveness. We will begin by surveying those immediate groups that are affected by the Authority's efforts, which may help in understanding the long-term support for the Authority's work. We will look at the formal publics and the informal publics, and focus especially on those voices that are critical of the Authority. We will also describe how anti-corruption activity began in Italy and in what ways it was supported or hindered.

7.1 Formal publics

The timeline of the Anti-corruption Authority's institution, described in Chapter 2, showed how efforts to address corruption by establishing an anti-corruption institution started in the executive branch in 2009. The acts passed since 2009 were intended to implement the Merida Convention of 2003, under the aegis of the UNCAC which bound all signatories to establish appropriate anti-corruption institutions. The domestic acts were complemented by a series of evaluation and reviews by international organizations. In 2009, the European Council's Group of Countries against Corruption (GRECO) published a document titled "Joint first and second evaluation round, evaluation report on Italy, 2 July 2009. 1st evaluation round: Independence, specialization and means available to national bodies engaged in the prevention and fight against corruption; Extent and scope of immunities. 2nd evaluation round: Proceeds of corruption; Public administration and corruption; Legal persons and corruption." In subsequent years GRECO published further reports on Italy, as well as other countries: "Compliance report on Italy, 23–27 May 2011"; "Addendum to the compliance report on Italy,

17–21 June 2013"; "Third evaluation round, evaluation report on Italy, Theme I: Incriminations; Theme II: Transparency of party funding; 20–23 March 2012"; "Compliance report on Italy, 16–20 June 2014". Then came the OECD Integrity Review of Italy in 2013, specifically requiring the country to establish an anti-corruption authority. In 2013 a decree was issued by the President of the Republic establishing a code of ethics for all public administration employees. New statutes were proposed by the executive and passed by parliament in 2014 which created the Authority as we know it today.

It may be observed how the dynamics of this process involved the highest levels of formal institutional stakeholders: international institutions, parliament and the country's executive. Efforts cut across political lines. Albeit with different emphases, the whole political spectrum was involved. An appreciation of the relevance of implementation emerges from the repeated efforts to reinforce the anti-corruption institution.

Events in Italy pre-dating these initiatives are worth referencing. The country's constitution, passed in 1948, establishes that international treaties have constitutional status, giving them "supranational" importance and power. As stated, Italy was among the six countries that established the basis for the EU, in 1957. And, of course, the country is also a member of the UN. It can be observed how the legacy of previous generations persists in current statutes and institutions. International organizations act as a virtual "fourth power" within each country and entreat each country to be responsible to the international community for their domestic behaviour. This point serves as an example on a global level, as there are many international regional organizations similar to the EU which can be brought to bear on domestic affairs in virtually all countries of the world.

In summary, the current formal stakeholders of the Authority are:

- Parliament, as the addressee of the Authority's formal reporting;
- The cabinet, which appointed the Authority's members;
- International organizations and fora such as GRECO, the OECD, the IMF, UNCAC and the EU, all of whom sponsored the establishment in Italy of an anti-corruption authority;
- Top management and all politicians who lead public administration organizations and public-sector companies, as the key addressees of the Authority's prescriptions;
- The trade unions;
- The Corruption Prevention Officers, as a key subset of public employees;
- All employees of public administration.

Parliament and the cabinet—despite having a different institutional position in relation to the Authority—appear to support the Authority and continue to request it to expand its activities, as we have seen in Chapters 2 and 3. As we have also seen, this attitude may have negative implications for the Authority's focus and effectiveness. Parliament and the cabinet, however, do not appear to be responsive to the Authority's recommendations.

International organizations sought and supported the establishment of the Authority via a variety of documents and binding protocols, such as the Merida Convention of 2003. The OECD WGB ("Working Group on Bribery") (2014) was important for guidance, support and benchmarking, although this contribution is strictly limited to a judicial perspective.

Public managers represent the core of the Authority's action, being addressees of the Authority's dialogue with public administration organizations. There are approximately 10,000 managers in the executive branch and a similar number in regional and local government. The Authority is aware that these managers may perceive its prescriptions and requirements as mere paperwork, a further burden on their workload. It is engaged in addressing these perceptions and is attempting to make its prescriptions useful and relevant in the day-to-day execution of their work. The issue of perception is explicitly addressed in the Authority's documents, as is its collaborative approach.

The trade unions of public managers and public employees have adopted a defensive approach and are attempting to minimize the significance of the Authority's role. They appear to not have grasped the scale of the task in hand.

The managers responsible for the Three-Year Plans to Prevent Corruption (the Officers) are a key subset of public employees, and their demographics and opinions are useful in helping to understand the Authority. Officers have a personal stake in the Authority's action: one of them even went as far as to ask to be employed by the Authority instead.

The key stakeholder of the Authority's activities is the individual public administration employee. All public administration employees are directly involved and are identified by the Authority as part of a class of motivated and independent civil servants. The Authority President quoted an IMF document on corruption at the second national meeting of the Officers on 24 May 2016:

> While designing and implementing an anti-corruption strategy requires change on many different levels, the Fund's own experience in assisting member countries suggests that several elements need to be given priority. These include transparency, rule of law, and economic reform policies designed to eliminate excessive regulation. Perhaps most importantly, however, addressing corruption requires effective institutions. While building institutions is a complex and time consuming exercise that involves a number of intangible elements that may seem beyond the reach of government policy, the objective is clear: the development

of a competent civil service that takes pride in being independent of both private influence and public interference.

(IMF, 2016)

7.2 Informal publics

The Authority's informal publics may include public employees (and their trade unions) and business. We began to deal with public employees in the previous section, on the subject of public managers, but if we look at the wider range of employees, out of a total 3.3 million employees, about 2 million are unionized, constituting 80% of the total 2.5 million unionized workers in Italy (recall that the total employed population of Italy is over 23 million). The Authority's actions are seen by the public administration trade unions as "management action", whereas it is the trade unions' objective to protect the "power entrusted with the individual" and "private use of public functions", which are actually part of the definition of corruption and which appear to include—in a broad interpretation of the terms—the tenured status of their employment contract and discretion in the discharge of their duties.

On the business side of things, the building contractors' association (ANCE) stands out as having the most dealings with the Authority. Italy's capital expenditure is 1% of GDP higher than Germany's (Cottarelli, 2016). Attention in this particular direction is also a function of the Authority's organizational heritage—as we have seen in Chapter 3—with a large proportion of its activities originating from supervision of public procurement of public works and infrastructure. This emphasis will only be reinforced with the new code of public procurement, which will broaden Authority activities in this field.

One of the Authority's strategic tactics was to emphasize the *cost* of corruption in order to engage businesses. However, this might be a difficult ploy as it runs counter to the logic of the corrupt status quo: those who are part of the mechanism have little incentive to collaborate with anti-corruption measures.

So far, we have looked at those stakeholder groups that not only have an interest in the Authority's activities but also have enough cohesion to aggregate and be vocal about them. But what about the stakeholders who may not even realize they have a stake, or those that are aware but are not able to organize and make their voices heard—the forgotten groups (Olson, 2008) and the latent stakeholders (Mitchell, Agle and Wood, 1997)? Industry in general, for instance, remains silent regarding the Authority, and the trade unions of workers in private companies are absent from the debate. Industry confines its activities in relation to public administration to industry-specific lobbying in a constant quest to relax regulation. Industry is worried about a short-term "freezing" of public administration activities as a consequence of corruption being tackled without a pathway having been found towards full-scale anti-corruption action—which is seen as a crucial long-term factor for economic development. Such stakeholder groups are potentially very large, using the figures we presented in Chapter 5 when discussing the economic context. Public administration has 3.3 million employees; approximately another million are included in the wider public sector to make a total around 4 million. We estimate there are another 4 million people employed in companies that supply the public sector, making a core of 8 million employed who are potentially involved in corruption. This leaves 15 million employed outside the corrupt equilibrium. Indeed, these are excluded from work by those in the inner circle of the government supply chain. We will refer to this majority proportion of the employed as the "unknown" stakeholders, who need protection from those whose mission it is to protect the people. Engagement of such groups could be a potential direction for Authority action.

7.3 The overt critics: magistrates and politicians

After the new Public Procurement Code (Legislative Decree no. 50 of 2016) was published, Authority guidelines gained more power. As well as non-binding guidelines or recommendations, the Authority also produces binding recommendations, and, wielding so-called "detail power", it can even produce executive orders. The Authority has opted for a low profile and moderate action and therefore chosen to send its guidelines to the State Counsel for their approval, although this is not mandated by law.

Considering this framework, the critics should be heard, as it is quite reasonable to be concerned that an independent authority—not subject to the people's control—has power over people's liberty. However, all Authority determinations or provisions can be challenged in the Administrative Court, a form of litigation that the Authority refers to as "jurisdictional litigation", i.e. that which the Authority is undergoing with regard to its own powers and other matters that are being challenged in court.

Among powerful critics of the Authority, an influential group of the magistrates of the judiciary are concerned that the Authority is too lenient with public administration; they also claim it is merely an instrument of the executive to improve its image. The only effective anti-corruption activity, it is claimed, is penal jurisdiction: repression and deterrence. This view sees a strong relationship between the state and state functionaries. Responsibility lies with each individual, with penalties for transgression.

As far as private companies are concerned, Legislative Decree no. 231 of 2001 involves new responsibilities for private organizations (corporations) and their top managers, in corruption and other penal cases. The innovation in law 231 was that not only would individuals involved in corruption be prosecuted, but the corporations themselves (and their CEOs) would also be liable. This law has been implemented quite often by the judges, but its significance and impact is as yet unknown.

From another perspective, a hypothesis can be drawn from Authority activities that penal action against individuals may encounter difficulties when crimes are committed within organizations: individual behaviour can be subsumed within an organization. Also, any deterrent effect is muted by an organization's informal sense of self-protection. Metaphorically, an organization is like Plato's Ring of Gyges, giving the individual a sense of immunity. An example is a case of corruption at the National Road Construction and Maintenance Company (ANAS), where seven managers and functionaries were arrested on 22 October 2015. The activity had been known about within that organization's headquarters and had been flagged up several times, but it took about 20 years for penal action to take place and remove the alleged perpetrators. (The jury is still out.)

The Authority premise about prevention in favour of repression is not intended to disregard and substitute for the magistrates' repression perspective. It seeks to be complementary to it, which is probably what the Authority's President was implying when he said, in the Milan Expo 2015 press release, "without taking anything away from previous forms of intervention". Moreover, it should be admitted that the Authority would be effective in its prevention efforts only in the long run, after a sustained effort.

Having said that, when speaking about a publicly controlled company that was part of the Milan Expo system, Ilda Boccassini, a prominent prosecutor in Milan, said: "We witness here a public entity which does not publish calls for tenders, anyway this is an issue for politics to take care of" (also in the press release). Three months later, on 12 October 2016, the public entity involved in the scandal (Fiera Milano SpA, owner of Nolostand, entirely involved with Milan Expo 2015) was in extraordinary administration imposed by the same judges, which implies the judges' appointment of a caretaker top manager—an administrator. The judges said: "This measure does not seek to be a sanction nor an act of repression, but it seeks to create a healthy environment of entrepreneurship in this area of the Milan Expo system" (*Il Giornale*, 2016: 9). Such opinions and lenient action are somewhat at odds with the premise of judicial repression and deterrence.

Listening to the critics is a useful exercise as it leads us to compare Authority action with judicial action: a comparison that cannot be avoided and should be tackled head-on. As we have already noted, we can think of the Authority as having a view of the whole economic system, a management view, whereas the judiciary has an "absolute" (or "specific case") view. A management view accepts trade-offs and works on a cost–benefit basis. The judicial view, in turn, could be described as a case-specific approach which does not consider the weight and relevance of a case in context. Judicial action is governed by law, and does not take into account considerations of cost and benefit; it does not make trade-offs. However, legal action also relies on a consequent deterrence effect over the rest of the economy and society, and such an effect is not deterministic: it is governed by behaviour that is studied by sociology and economics. So the resources of economics and sociology are, perhaps implicitly, also availed of by the judicial branch. The stakeholder landscape we have seen so far includes supporters of the Authority and detractors. The detractors might be more numerous than we have so far suggested, to the extent that a politician has seen fit to try to give a voice to latent adversarial stakeholders. Political leader Stefano Parisi (former city manager of Milan) in November 2016 declared:

> the Prime Minister delivered Italy into the hands of the magistrates, starting with the Authority president, who is a great illness for our country. Who is he, what is his institutional basis? The Authority is not mentioned in the organization of the Italian state and it creates a lot of confusion within public administration.
>
> (*Repubblica*, 2016)

The newspaper article said the Authority's President was a "curse" for the country.

There is also a strand of criticism towards the Authority from administrative law scholars, who disagree with the President of the Authority being given a judicial and a public prosecutor's role. As stated, they see the Authority as a "big policeman" who is trying to "moralize" the public sector. The specific case studies the Authority has been publishing are seen as a breach of the independence of public organizations.

In the following chapter we will present some of the on-the-ground situations that have given Italy such a bad reputation for corruption.

Chapter 8

Corruption case histories

8.1 An anti-corruption "clinic"

However neat the design of the implementation process may appear, in real life things happen one after another in a micro-sequence: phone calls do not each come with a tag stating their importance towards meeting the over-arching goal of the general long-term welfare of the country and of its economy. With this in mind, it is instructive to describe the Authority's day-to-day work.

The authority also acts as a "clinic", holding informal support meetings with organization representatives regarding their task of writing an anti-corruption plan. Some organizations are taking it seriously, behaving diligently at least, and are requesting meetings with the Authority. Examples are:

- The national TV concern, RAI (9,000 personnel; turnover: €9 billion);
- The Municipality of Rome, which we have already met (26,000 personnel, of whom 6,000 are in the police corps; €6 billion budget; plus in-house and companies participated with for municipal services: power, sanitation, mass transit);
- The central bank, Banca d'Italia (7,000 personnel; a central bank, formally a shareholder corporation).

8.2 A large municipality

A specific case in point happened when in 2014 the Municipality of Rome was beset with scandals, leading to public outrage. These broke out in at least three areas of municipal activity:

- Municipal police supervision of local commercial activities, such as restaurants—extortion;
- The existence of a shared speed-money "bribe price list" for building permits—this list was used by the municipal police and by the technical staff of the municipal districts; and

- Bid rigging and absence of tenders also took place in procurement of services.

The city's budget for external contracts was allocated on the basis of an agreed formula: three-quarters to companies supported by the political majority and one-quarter to companies supported by the opposition. This scandal was dubbed in the media "Capital Mafia" (*Mafia Capitale*), in a reference to Rome. This took place both in the procurement of social services to non-profit companies, mostly cooperatives, and in the procurement of waste disposal services. There was no tendering process: services were contracted and awarded directly. A network of politicians, managers and cooperatives was uncovered that had been managing overtime contracts with a total budget in the order of tens of millions of euro—out of a total city budget of €6 billion, to provide some context of the scale of the operation. Corruption here took the form of a network of favours and the bypass of due process. Professor Giarda, the country's first spending-review czar, said in a 2012 paper that it is very likely that the value of the contracts was above the market rate.

Scandals notwithstanding, the Municipality of Rome's anti-corruption plan was largely insufficient. The plan went through two drafts: the first framed every action proposal in the future tense ("we will do"), but even the second remained insufficient. However, the chief of the municipal police corps, on his own initiative, decided to implement a measure that was only mentioned in the plan and not substantiated with any operational instructions: rotation of police across the city's 19 precincts (*gruppi territoriali*). This generated immediate trade union uproar among the Rome police corps and the police chief asked the Authority for assistance. The Municipality of Rome police corps comprises 6,000 personnel, of which 2,500 are (middle-level) functionaries (*funzionari*) and 3,500 agents. They have four specializations: traffic (*polizia stradale*), administrative supervision of local business (*amministrativa*), buildings and construction (*edilizia*), and socio-environmental (*socioambientale*). Of the 6,000 employees, an astonishing 1,000 are trade union representatives. The Authority ruled that trade union representatives are also subject to personnel rotation. Finally, the Authority admitted that a limit to rotation may derive from the "high technical content" (*elevato contenuto tecnico*) of certain jobs but that it did not apply here. It is very likely that these decisions will be challenged in the courts.

8.3 An international event

International attention was drawn to corruption in Italy during the years of preparation for the Expo 2015, to be held in Milan in the northern region of Lombardy. Site construction is the major operational undertaking of such global exhibitions, and construction implies large expenditure of public monies, managed through tenders. The Municipality of Milan set up an ad hoc corporation, the Expo Company, which was to manage the contracting and other essential tasks leading up to the opening of the exhibition in May 2015. Its budget was €15 billion (in the context of the €250 billion of annual government procurement expenditure [2011]). Through 2012–2014, several instances of corruption were exposed in both the Municipality and the Expo Company.

The Milan Expo 2015 contracting scandals were observed by the OECD, and the Authority intervened with a new organizational procedure that appears to have met with good results. The model was to appoint an extra-ordinary administrator—an ad hoc manager—for a specific contract, not necessarily for the whole contractor company. An ad hoc control unit oper-ates in real time on the site of the contracting unit, i.e. the unit of the public administration that issues the contracts. This "commissioneering" procedure is in contrast with the mainstream approach where a judge intervenes, halts all activities related to the specific contract, and often also suspends all activities of the company under investigation. Under the new approach, work in pro-gress does not come to a halt but a control mechanism is coupled to the ordinary progress of work. We have seen above that the continuation of work in progress and the avoidance of administrative freezing of operations is the major concern of industry when dealing with corruption issues and gov-ernment intervention.

This case is an interesting example of balancing the cost of control with the benefit of control. There is a line of thinking in the area of organizational behaviour that sees control as a key element in getting an organization—and the people who work in it—to behave as the mission of the organization would require. However, those who hold such viewpoints are also aware that control does not come without cost, both on the input side and—mostly—on the output side. Therefore, the thesis on optimal organizational behaviour states that the cost of control should not exceed the benefit of control. The "commissioneering" of Expo 2015 was made necessary by the critical dead-line of 30 April 2015: Expo 2015 was due to open to the world on 1 May and failure of that to happen would have been an unbearable cost.

The Expo procedure was the result of a specific agreement between the OECD, the Expo Corporation and the Authority. It may be of interest to read the press release:

> This afternoon in Milan the Protocol of Understanding for the fight against corruption was presented. The company of Expo Milano 2015 has made the agreement with the Italian National Anti-corruption Authority (ANAC) and the Organization for Economic Co-operation and Development (OECD). The press conference was held at the end of the first official advisory board meeting held by the OECD at the head-quarters of Expo Milano 2015 and was attended by Raffaele Cantone, President of the Authority, Nicoletta Parisi, a member of the Authority

Council, Francesco Paolo Tronca, Prefect of Milan, Giuseppe Sala, Sole Commissioner for the Government for Expo Milano 2015, Nicola Bonucci, Director of Legal Affairs of the OECD and representative organization in international posts, Antonio Capobianco, senior expert of the OECD Competition Division, and Paulo Magina, director of the Public Procurement Division on integrity of the public sector of the OECD.

The putting in place of the anti-corruption plan developed by Expo Milano 2015 in collaboration with ANAC and OECD [...] creates a reversal of perspective, based on the full transparency of all procedures from their early stages: "Without taking anything away from previous forms of intervention", said Raffaele Cantone, "the preliminary verification of the correctness of each document makes detection of irregularities faster and more efficient. It is also an important issue to ensure that work is progressing according to its outlined schedule." The mechanisms of control over the regulation of the records remain active and are carried out in a continuous dialogue between Expo Milano 2015 and ANAC. There is an attitude of appreciation [that Expo Milano 2015 succeeded in opening on time] from the Prefect Francesco Paolo Tronca: "The cooperation of Expo Milano 2015 is crucial for both types of control that are carried out in accordance with the enforcement of the law, whether from the estimates received of information stored in the database or that work which is carried out directly on construction sites."

(Expo Milano 2015, 2015)

The Anti-corruption Plan for Expo Milano 2015 has the potential to be a workable model internationally. As Nicola Bonucci explained:

The involvement of the OECD in the anti-corruption protocol of Expo Milano 2015 is particularly important, because it allows us to benefit from the extensive experience gained in the fight against corruption acquired both from other countries and the OECD itself. Our job is to collect and compare the best anti-corruption strategies put in place by the various countries, with the aim of identifying the most effective, and not repeat the same mistakes. The Expo model is particularly significant for us and it is not inconceivable that it could become a reference point at international level, applicable to other major events.

(Expo Milano 2015, 2015)

We should also note that judicial comments about "codes of ethics and memoranda of understanding" were probably referring to this memorandum. It may be worth mentioning that the Expo 2015 Commissioner Giuseppe Sala went on to have a political career and was elected mayor of Milan on 19 June 2016, a sign of popular approval of the handling of the Expo initiative in its final stages.

8.4 The Ministry of Public Works

In mid-March 2015, engineer (*ingegner*) Ercole Incalza was arrested by the judiciary on charges of illicit conduct in his function as head of a Mission Oriented Unit, an ad hoc organizational unit of the Ministry of Infrastructure (formerly the Ministry of Public Works and Transportation, where the accused had been prominent since the 1980s). Being over 70 years of age, Incalza was granted house arrest. As in the Expo case, a major concern was that judicial action should not halt the continued functioning of the unit. Parallel to the Expo case, a memorandum of understanding was signed between the Minister of Infrastructure, Pier Carlo Padoan, and the Authority. As of the end of 2017, Incalza had been acquitted 16 times of all charges brought against him.

8.5 Conclusion

Specific cases about Italian public administration have caught the attention of international media. In 2014, *The Economist* reported on the high salaries of the employees of the Italian parliament: "To earn €136,000 ($181,590), a browse of the internet suggests, you need to be an IT operations director at a British firm, governor of New York state—or an *usciere* (usher) in the Italian parliament." It further pointed out that "there are 11 unions who have members working in the administration of the lower house and 14 in the administration of the Senate" (*The Economist*, 2014a). Two years later, commenting on the Italian public administration, the same magazine said: "Regions and municipalities are the most corrupt layers of government." Italy "would have been better off arguing for more structural reforms on everything from reforming the slothful judiciary to improving the ponderous education system" (*The Economist*, 2016c).

However, these cases are only part of a global framework in which underperformance in public administration is the rule rather than the exception. In North Africa: "The inefficiency of Arab bureaucracies matters to more than just frustrated citizens. The uprisings of 2010 that toppled regimes in the so-called Arab Spring were as much a cry for services as for democracy" (*The Economist*, 2015; see also World Bank, 2015). In Latin America:

> The plodding inefficiency and red tape of public bureaucracies has become an unaffordable drag on the region and a source of growing frustration. Despite the economic slowdown, more Latin Americans are middle class than in the past. They are demanding a more sophisticated, efficient and less corrupt state. Decentralization and the digital revolution pose additional challenges [...] The region's civil services suffer many vices. One is an obsession with procedures and hierarchies and a disdain for service and outcomes. Many Latin American civil servants must

follow thick procedural codes but are not made accountable through performance targets. Organs of control fail to prevent corruption but instil a terror of initiative.

(*The Economist*, 2016a)

Europe fares no better in the appraisals:

The French call them *hauts fonctionnaires*, the Germans *Beamte im höheren Dienst* and the British, somewhat more economically, know them as "mandarins". The senior echelons of civil services are a powerful arm of the state. They implement the reforms dreamed up by politicians, and design public services ranging from welfare systems to prisons. Compared with private-sector bosses, the bureaucrats who manage the public sector tend to be less well paid but have more cushioned lives, with more secure jobs and far less pressure to improve productivity. Now the mandarins face change. [...] Too many civil servants, especially in continental Europe, swirl around a bureaucratic Gormenghast but rarely leave it. Nearly four-fifths of German senior public servants have been in public administration for more than two decades. The French state under François Hollande is governed by a caste of unsackable functionaries, resistant to reform.

(*The Economist*, 2014b)

We will end this chapter by reinforcing the point that, when corruption is defined in its wider meaning of maladministration, we are looking at a mass phenomenon which is "sub-criminal", yet nonetheless with strongly negative implications for the efficiency and effectiveness of public administration and democracy in general.

Chapter 9

Joining public and private anti-corruption efforts

Chapters 5–8 have aimed to broaden the spectrum of what must be taken into account in anti-corruption activities: Chapter 5 covered the economic and social context of the anti-corruption effort; Chapter 6 looked at how anti-corruption efforts can be measured on all levels, intermediate, outcome and opinion; Chapter 7 discussed the process of setting the agenda for a sustained anti-corruption effort; and Chapter 8 cited specific examples of potential corruption scenarios.

In Part I, the anti-corruption plan emerged as a potential instrument for documenting corruption and anti-corruption efforts and encouraging international dialogue. Based on the observations made in Part II thus far, it can be asserted that anti-corruption metrics need to be integrated with those of their organizational environment. A need would seem to arise for measurements to evaluate anti-corruption activity in order to ascertain whether it is focusing on important issues in the organization rather than marginal ones. The "materiality" of the issues and of the metrics should be assessed, and we should see measurements of "denominators", i.e. to evaluate the sum totals of the anti-corruption efforts. Such requirements are already evident in current anti-corruption plans in risk analysis reports.

But in considering the implications of such risk analysis, we realize that what we are calling for is an all-encompassing report on the whole organization, without which the risk is run of being fully compliant with anti-corruption measures at the expense of the organization's core mission. In fact, there is no instrument nor provision in public administration that explicitly calls for the fulfilment of the mission the organization was established to pursue. Administrative control and judicial review are mostly about alleged negative behaviour; they rarely concern the omission of positive action. This is why the foregoing chapters have called for an expansion of the anti-corruption plan.

9.1 The convergence of reporting instruments

Considering the premises developed in this book thus far, a need is indicated for the integration of the anti-corruption plan with information about the organization's financial performance and its "product" or service—the latter being something that must be defined in the plan. The notion of "product" might be odd for a public administration, but it has a quite clear definition in the context of managerial control in non-profit organizations (Anthony and Herzlinger, 1975; see also La Noce 2015 for a practical example).

With such information included, the anti-corruption plan would assume the nature of an institutional report that covers all aspects of the organization but with a particular focus on corruption containment. Such a report might be quite similar to a CSR or sustainability report, which in fact may include sections on corruption containment and codes of ethics and how these are implemented within the organization. The Corporate Responsibility Summary and the 2016 Annual Report of the British defence concern BAE Systems are an example that can be considered paradigmatic of these instruments in the private sector.[1]

The private sector is also a rich source of guidelines for organizational and risk analyses, with all the relevant documents. The Global Reporting Initiative (GRI) is one such instrument, producing standards and providing follow-up and benchmarking of different corporations.[2] Integrated Reporting is another, under the aegis of the International Integrated Reporting Council (IIRC).[3]

Integrated Reporting in public administration, then, will be an expansion of the corruption prevention plan to place anti-corruption activities within the organization's core mission and the activities that are carried out to fulfil this mission. To capture the mission of the organization and be accountable for it, such a report should be integrated with financial reporting, because the latter provides a 360-degree view of the organization. The integrated report will include contextual description and how the organization fits within this context; it will include production metrics and descriptions of stakeholder relationships. These are subjects discussed in Chapters 5 (context) and 7 (stakeholders).

A public administration social responsibility report should include specifics that the GRI outlined in its Public Sector Supplement to its general guidelines. The primary consideration appears to be that a public administration organization's economic bottom line ought to be more rigorous than that of a private company. This apparent paradox is because of the absence of mission- and output-specific content in public administrations' financial statements as they currently stand. Public financial statements, in fact, only certify the legitimacy of the administration's actions rather than explaining their activities' relevance to their core mission. By contrast, an organizational social responsibility report considers the quantity and quality of the work undertaken by the organization—which should be, in essence, the economic bottom line of government, i.e. the place where all government action is accounted for in the most rigorous way.

Similarity implies convergence. The observed similarity between reporting tools in the public and private sectors, and the original provenance of such tools, should not surprise us, as all organizations are organizations first and foremost, and the need for reporting is primarily a consequence of their organizational nature rather than their public or private status. The tools of anti-corruption initiatives and the reporting tools of such initiatives are therefore observed to be converging towards a unified model. We can expect to observe the same phenomenon in examining the initiatives behind or accompanying such activities, as will be shown in the next section.

9.2 The convergence of United Nations initiatives

The UN Convention Against Corruption (UNCAC) and the UN Global Compact (UNGC) speak to one another. The former called for the creation of ad hoc public organizations to contain corruption, and the latter echoed this concern by calling on private companies to play their part in containing corruption, alongside their efforts on human rights.

> The tenth principle against corruption was adopted in 2004 and commits United Nations Global Compact participants not only to avoid bribery, extortion and other forms of corruption, but also to proactively develop policies and concrete programmes to address corruption internally and within their supply chains. Companies are also challenged to work collectively and join civil society, the United Nations and governments to realize a more transparent global economy [...] With the entry into force of the United Nations Convention Against Corruption in 2005, an important global tool to fight corruption was introduced. The United Nations Convention Against Corruption is the underlying legal instrument for the 10th Principle—on corruption control—of the United Nations Global Compact. Corruption can take many forms that vary in degree from the minor use of influence to institutionalized bribery. Transparency International's definition of corruption is "the abuse of entrusted power for private gain". This can mean not only financial gain but also non-financial advantages.
>
> (UNGC, Principle Ten: Anti-corruption)

Note that the UNGC is being implemented internationally through a recruiting mechanism to enrol companies into the programme, which accounts for about 9,000 corporations worldwide in 2018.

The UNCAC's call for countries to set up anti-corruption authorities for integrity in public administration is being monitored by the UN Office on Drugs and Crime, administered through a remarkable questionnaire procedure aimed at anti-corruption organizations.

The UN Office on Drugs and Crime is the body that originated and now manages the UNCAC. Its task with regards UNCAC implementation is to conduct a review cycle with the objective of verifying each country's implementation effort and their effectiveness in containing corruption as a result of the Convention. The "mechanism" looks very similar to the GRECO review rounds. As regards Italy's UNCAC implementation, the review within the UN Office review cycle is being conducted by the US and Sierra Leone, which, as of 2018, were compiling a "Country Review Report: Italy/Review by the United States and Sierra Leone of the Implementation by Italy of Chapter II (articles 5–14) and Chapter V (articles 51–9) of the United Nations Convention Against Corruption for the Review Cycle 2016–2021". The draft of this was very detailed, a document of more than 400 pages. It took the form of a questionnaire, the administrators of which required an emphasis on practices rather than legislation, i.e. they took an implementation approach. The review asked whether there was "active involvement of the media and society in social accountability such as the popularization, monitoring and implementation of the Convention"; furthermore, they stipulated that information on society's participation in developing (corruption containment) policies was required to determine Italy's full implementation of (anti-corruption) provisions.

Both systems have their limits, however. A shortcoming of the UNCAC as it currently stands may be the implementation process: a shortage of international cohesion and benchmarking among the initiatives the various countries have taken to implement the convention. Potential limits of current UNGC implementation (on corruption) are its exclusion of public administration and its possibly divisive message, i.e. fostering a public vs. private mentality, targeted as it is solely on the private sector.

There are further instruments and relevant messages from other sources. In 2015, the UN approved the 2030 Sustainable Development Goals (SDGs),[4] which include Objective 16.5: "Substantially reduce corruption and bribery in all its forms." This is a subsection of Goal 16: "Promote peaceful and inclusive societies for sustainable development, provide access to justice for all and build effective, accountable and inclusive institutions at all levels." The UN Sustainable Development Solutions Network (SDSN)[5] presents Proposed Indicators to address the aforementioned Objective 16.5: "91. [sic] Revenues, expenditures, and financing of all central government entities are presented on a gross basis in public budget documentation and authorized by the legislature; 94. [sic] Perception of public sector corruption."[6]

The EU in turn intervened with a 2014 directive, no. 95, mandating disclosure of non-financial information on the part of large corporations. Member states have been ratifying this directive, and the first results are expected from corporations by 2018 with the publication of non-financial reports.

Once we pivot on the plan, dialogue becomes possible. Indeed, synergy may happen. That is the subject of the next section.

9.3 Finding public–private synergies: a proposal

The private sector is at least more visible than the public sector in its initiatives to promote transparency and anti-corruption. One key factor of this visibility is the centrality of reporting, which we propose can be leveraged to work for public organizations as well.

This provocative proposal, which challenges constitutional subject matter, would require public administration to publish anti-corruption reports and plans and subscribe to the GRI, so that individual public organizations can be open to international review. This would imply an engagement with the UNGC Principle #10 on corruption and the UNCAC. It is reasonable to expect UN organizations to converge towards the UN 2030 SDGs, especially Goal 16 ("Promote just, peaceful and inclusive societies") for which an anti-corruption effort is crucial.

Our proposal is to make the anti-corruption plans a fully fledged reporting tool and accelerate the process of convergence of public and private efforts towards joint "organizational" social responsibility, overcoming the sometimes-impeding distinction between public and private sectors.

A comparison mechanism should be activated at an international level. Each public organization could publish their anti-corruption report on those international websites that have been established in the wake of the Global Compact initiative, e.g. that of the GRI.

Nor should our thinking be distracted by the consideration that we are dealing with public administration which is intrinsically weak and short of funds. Many cities perceive themselves as international competitors and manage budgets in the range of billions of euro. It could therefore be beneficial to establish a self-sustaining mechanism of reporting and feedback about their social responsibility. Let us be aware that European rules require private companies with a turnover of over €20 million to comply with non-financial reporting. A medium-sized municipality in Italy has that kind of budget—and there are thousands of them.

This proposal has the additional virtue of streamlining anti-corruption reports, making them less of a compliance effort and more of a substantive effort, in alignment with private-sector initiatives.

This represents a challenge and is not something that can be achieved in the short term. The rationale for making a proposal such as this is that easily attainable mandates are frozen into compliance routines, thus frustrating their original purpose of improving society and the economy. They even have the effect of stymieing the efforts of those who voluntarily embark on innovative activities.

The consequences of applying CSR to public administration would result in the following beneficial activities:

• Voluntary managerial evaluation, research and disclosure of results—this is basically part of the major fields of public policy and public management studies which are used as routine managerial tools in public administration;

- Peer review and customer satisfaction exercises;
- Comparative international analysis of efficiency and effectiveness;
- Awareness and adoption of Integrated Reporting (Eccles, 2010; IIRC's Public Sector Pioneer Network[7])—our view is that Integrated Reporting and CSR are close, if not overlapping, domains;
- Emphasis on impact and outcome rather than input and expenditure; and
- Assessment of the competitive environment wherein each public administration operates.

Our proposal to give full recognition and international visibility to anti-corruption reporting and plans does in fact sit comfortably within the existing cycle of planning and feedback. The very notion of transparency and the emphasis on publication of reports imply an aspiration towards a virtuous cycle of improvement via interaction with potential publics and welcoming criticism. This proposal asks merely that we enhance this virtuous cycle through global evidence of public organizations' anti-corruption efforts and activities.

9.4 Beyond control: management by benchmarking

Checks and balances within public administration have assumed the shape of further controls and regulations, not the shape of a self-regulating mechanism, which is what the actual notion of checks and balances implies: weight and counterweight. Checks and balances in public administration look non-dynamic and inflexible. They have been leading to increasingly less discretionary power, and yet such a lower level of discretion has accordingly led to more arbitrariness in the execution of this lower discretion. Irresponsibility and stasis could be the result as the threat of judicial review hangs above them. On the other hand, there is an abundance of reliance on the virtues of public administrators: codes of ethics, calls to ethics, even an Office of Government Ethics. As such, we entertain the hope that in future the model of organizational behaviour for public administrations—which is implicit in the law—will take into account those organizational behaviour theories that have been formulated over the past century which describe and modify the Weberian ideal of rational bureaucracy. In this book we have offered a specific way of integrating such theories and evidence into the reality of anti-corruption efforts.

Notes

1 See https://investors.baesystems.com/agm for the 2017 reports.
2 www.globalreporting.org.

3 https://integratedreporting.org.
4 https://sustainabledevelopment.un.org.
5 Launched by UN Secretary General, August 2012; unsdsn.org.
6 http://indicators.report/targets/16–5/.
7 "Spreading the awareness of Integrated Reporting in the public sector". See http://
   integratedreporting.org/ir-networks/public-sector-pioneer-network/.

Chapter 10

Conclusions and future studies

In this book we have aimed to gain an understanding of the operating concepts of anti-corruption efforts in several countries, following the UNCAC. Part I discussed legal and organizational concerns, of which Chapter 1 was an overview of the anti-corruption efforts of several countries, Chapters 2 and 3 looked at Italy's experience, and Chapter 4 formed generalizations by comparing several countries' efforts. Part II was concerned with socio-economic factors, and its chapters (5–8) dealt with the socio-economic context, measures of corruption, stakeholders and specific instances of corruption. Finally, Chapter 9 nailed our colours to the mast and brought forth a proposal for integrating public and private efforts via the public–private joint reporting platform of the Global Reporting Initiative.

The core of this book analyses one specific anti-corruption organization in detail, and it is this specific example that underpins our global perspective. There is relevance to both developed and developing countries because the overview of developed countries is followed by a close inspection of a country that has as much in common with the global south as it does with the global north. The book bridges anti-corruption efforts between the public and the private sectors, and offers an administrative law perspective as well as a management one.

This book follows an implementation approach, looking both at the normative reality of anti-corruption organizations and at the actual output and outcome of such organizations. The idea was to trace the impact that anti-corruption organizations have on the economies and societies they are trying to affect. Such an implementation approach brings legal administrative and management studies together, making the book as relevant to administrative law as it is to CSR, fields that are relatively distant from one another.

This study has approached the general issue of corporate and organizational social responsibility in public administration from sideways on. In fact, anti-corruption plans can be seen as CSR reports, taking an outside-in view of public administration, i.e. public administration looking at itself and disclosing its own workings. This finding was leveraged to put forth our proposal that public administration organizations enter the global arena of CSR reporting *en masse*, accepting the challenge of comparing themselves with their global counterparts. The institutional environment of this international arena is defined by several UN initiatives: the UN Global Compact (UNGC), the UNCAC and the UN Sustainable Development Goals (SDGs).

We will make a few further comments before suggesting routes for future studies.

The anti-corruption efforts place specific emphasis and focus their attention on public-sector behaviour and impact on the economy. This is probably a novelty in the cultural and administrative landscape of many countries, where judicial emphasis on corruption has traditionally been devoted to private-sector operators.

In the international arena, the concept of corruption prevention does not loom large in the literature, where corruption case histories are predominantly about corruption repression. The anti-corruption organizations reviewed here, however, demonstrate the newer approach.

The idea of organizational change as a means of preventing corruption, although a very general one, seems conceptually sound. Again, it is scarce in the literature, whereas many instances have been identified in this study. For instance, such receptiveness to change is evidenced when the Italian Anti-corruption Authority feeds back into the legislative process to make rules more effective. This points to an implicit notion that corruption cannot be contained within the boundaries of current legislation, norms and organizational arrangements; it is not eradicated by a stricter implementation of current norms. If corruption exists, it must in some part be attributed to ineffective rules. Corruption is contained via change management and new organizational arrangements.

The anti-corruption experience is one of attempted substantive dialogue between the "centre" and the branches of government, i.e. the local units of central government and the units of local and regional government. For instance, the Italian Anti-corruption Authority (which can be considered part of the "centre") worked on the Three-Year Plans to Prevent Corruption (written by the branches of government) and provided feedback to the diverse organizations of public administration. An actual feedback cycle was established.

Most of the anti-corruption organizations examined here have within their field of vision all the parts of economy that might possibly be involved in corruption. Through such a comprehensive approach, these organizations have drawn sustained attention to the issue of corruption and are putting pressure both on their governments and on public opinion to tackle the issue. Such attention, though productive in terms of activities and law-making, may also bring to light the difficulty of capturing such a complex and pervasive phenomenon.

With regard to possible future studies, the first and most obvious work to be done is the continuation and the extension of this narrative effort to all of the organizations that have been reviewed here. This is the first of a number of needed comparative administration studies. Over the course of time, the anti-corruption organizations will deploy new actions that should be recorded for the benefit of an international audience. And public administrations will respond to their anti-corruption organizations, which represents a major activity demanding documentation. In fact, recording and documentation is an essential part of the whole process, if it is to be worthwhile.

Criticism and evaluation of the anti-corruption experience will also be needed. A convergence trend can thereby be sought, as metrics may become more specific and help in detecting indications for corruption containment. On this subject, the Italian Authority has been assigned funds from the EU to conduct research into identification of objective indicators for corruption and of corruption containment.

The relevance of the public administration perimeter is a key finding of this study. Comparison between public administrations of different countries is rare and seldom quantitative, and, moreover, organizational units and sectors that have the same name might be different when we examine their organizational arrangements. So it is critical to clarify and quantify what is meant by "government" (or "public administration") in each country of the world and to ascertain how each sector is organized. This needs to be done in order to shed some light on the relatively unknown side of CSR: public administration. This is the second comparative study presented here.

Future studies may also find a link between judicial effectiveness and the relative effectiveness of civil law and its feedback into the social fabric.

The detail provided in this book may be of use in investigating the new separation of powers. It could further our understanding of which powers are necessary and distinct from Montesquieu's original three powers and hence require separate branches in the form of an independent authority. Which functions (counsel, administrative sanctions, guidelines) and which subjects (transparency, corruption, public procurement) can be performed by existing government agencies, and which ones require an additional branch and a discrete power? Under the logic of separation of powers, a new articulation of power might be useful in the form of an independent authority implementing the law without a political interpretation and without being subject to the political power of the cabinet.

Future studies may also adopt a global perspective by means of comparisons with other relevant situations. Such an approach might also address perceptions of uniqueness as an attitude common to those who are struggling to contain corruption. People across the world tend to perceive corruption—as they do many other issues—as local problems, specific to their culture, history and destiny. The good news is that corruption is a global problem. Everybody has it. Tolstoy's opening sentence in his novel *Anna Karenina* lends itself to a useful paraphrase. Whereas Tolstoy says, "All happy families resemble one another; but each unhappy family is unhappy in its own way", we could say that every country thinks it is corrupt in its own unique way, but that is not the case. We risk falling victim to the "Tolstoy trap" of thinking our problems define us as different, whereas we are in reality all affected by the same plague of corruption. This account of the experience of several countries may therefore be helpful at a global level, through further studies in comparative public administration.

Finally, identifying that there is a widespread definition of corruption as "maladministration", as has been discussed in this book, is a useful outcome in itself since it highlights the theme of "evasion of work". Evasion of work is not an activity that is subject to evaluation. It is an activity that may also be performed in good faith but, without evaluation, it remains merely an activity; it does not become work. "Work" can only be an activity performed by one party that a second party appreciates and is willing to pay for.

Appendix

Interview with the Italian Anti-corruption Authority President, Raffaele Cantone

On corruption in general

Q. What are we talking about when we speak of corruption? Are we talking about specific crimes or deviant behaviour on the part of public employees or are we talking about inefficient and "bad" administration?

A. Corruption is first of all a penal phenomenon, involving one public party and one private party. However, this is only one individual act of corruption. In a prevention perspective, corruption is a wider phenomenon; it encompasses all private-interest behaviour of public officials. So administrative corruption includes penal corruption and it is wider than penal corruption.

Q. How do you measure corruption? How important is it to have objective indicators to measure corruption risk or to measure how much a specific country is containing corruption?

A. I don't think that today any objective measures of corruption are possible. This is what the penal professions call the "dark number". We do not know this number because there is collusion between the selling party and the buying party; therefore, corruption and its quantity does not emerge. Information emerges only from a conflict of interest. In corruption there is no conflict of interest between the involved parties.

However, what are more important are indicators that can reveal potential corruption: "spy" indicators, markers of possible corruption. For instance, procurement carried out via imperfect public procurement systems could be a necessary, but not sufficient, condition of corruption. There is also a third category: indicators of prevention effort. These would be "non-negative" indicators, e.g. use of correct procurement procedures shows there is more respect for the rules. Where repression did work or prevention did work, procedures are more correct. You can measure a country's capability of dealing with corruption by its respect for the rules.

There is no objective system for measuring corruption. Our challenge is to find indicators of prevention of corruption, indicators of our containment effort. For instance, we do the Three-Year Plans to Prevent

Corruption; that is part of our effort. This is also important at the international level.

Q. We understand that by "dark number" you mean the sum total of bribes and kickbacks. The most accredited measure among scholars is "foregone" GDP, i.e. the national product that is lost because corruption ultimately makes society less effective and efficient from an economic point of view. Foregone GDP is much bigger of the sum total of bribes and it does not even take into account the "social moral cost" of being aware of living in a corrupt country.

A. Yes, they are. Today we have perception indices. They are useful, even though these indicators do not say anything about the quality of our containment effort. However, perception indicators do express the low trust that citizens have in public institutions. We also need to take into account that indicators of corruption perception might actually be augmented by our containment activity. In fact, a containment effort brings more public attention to corruption; such renewed attention, in turn, increases the public's perception of corruption.

The same is true for risk indicators because they give us an idea of the need—or the incentive—for people to seek corrupt ways to work with public administration.

So we need to work harder on such indicators. We started a special project on corruption indicators, sponsored by the European Union, involving several scholars who were also involved in the European "Antocorrp" Project.

Increases or decreases in the cost of infrastructure, for instance, we consider risk indicators. Risk indicators are also useful because they tell us the direction we need to go. They give us hints on how to work on containment of corruption. We could say they are "constructive" indicators.

Corruption in Italy

Q. Reading your book *The Italian Illness*, one might get the impression you think corruption is endemic, specific and deeply rooted in Italy. Is that so? Is there no hope for Italy?

A. Corruption in Italy is in many instances "systemic", i.e. all adjectives apply: there are specific situations, organizations or environments, where corruption is endemic and deeply rooted. However, this is not true for all administrative areas.

We still do not have an agreement, we are not socially aware of the systemic corruption that is present is in many administrative areas. We still need to understand the seriousness of corruption in Italy.

Q. Is Italy unique?

A. Not at all. Italy is very similar to many Latin countries. We share with those countries our Christian Catholic tradition and our relationship between the citizens and the public institutions. In Christian Protestant countries the citizens have a stronger relationship with their public institutions.

Our penal approach goes in the same direction. We do not care so much about the ethics of public officials. We do not do anything about anomalous behaviour that is not penal. A case in point is public employees cheating on their time clocks: after penal charges are dismissed, there is no civil follow-up; there is no issue of an ethical relationship between the cheaters and their colleagues. Once the penal trial is finished, everything is finished.

Q. Penal action nonetheless seems to be more lenient on public managers than on private managers.

A. In business firms, the entrepreneur is responsible. In public administration the individual manager is responsible. From a penal point of view, however, the prosecutors will include all who were instrumental to the crime. A public administration top manager favouring penal action, then, appears to be collaborating with justice and working on corruption containment.

Q. We have said that the perception of corruption is so high in Italy because there is little trust in public institutions. What is the specific character of corruption in Italy?

A. Penal corruption lost its dual character. It no longer derives from the collusion—the "same will"—of a public official selling corrupt services to a private party. Now buyer and seller are on the same side. Public officials are on the payroll of private enterprises. The relationship is today organizational rather than one-to-one, or dualistic.

Containing corruption: between repression and prevention

Q. After many years of strong emphasis on penal repression, today there is an emphasis on prevention of corruption in public administration. Are the two strategies at odds with each other or is there an effective equilibrium between them?

A. Both strategies are necessary. Prevention is about avoiding the occurrence of corruption. However, prevention is also necessary in the rehabilitation of a corrupt environment. Prevention provides the tools for negative situations to come back on track.

Q. The judicial branch does not appear to think much of administrative prevention of corruption. What is the idea behind this opinion: on the one hand only penal repression can eradicate corruption or on the other hand there is a conviction that administrative prevention is so complex and difficult that it cannot deliver within a reasonable time-frame? Or is there a third opinion whereby administrative authorities should only work to detect and report corruption to the judiciary?

A. It is true: many in the judiciary think prevention does not help. However, a progressive part of the judiciary think repression alone is not enough. The low trust of the judiciary in corruption prevention is due to historical and cultural reasons we have already explained. This is also due to previous attempts at prevention that did not work. Mistrust does have some objective basis.

There is a third position: prevention should be analogous to the police—reporting to the judiciary. On the other hand, I believe prevention is like workplace safety: it is meant to prevent hazards and casualties. Then it may also call on the police or the judiciary when it identifies environments that are systemically corrupt. Prevention is young in Italy and it has not yet developed its own specific autonomy.

Q. Several times you said prevention of corruption is about objective organizational measures, against corruption, and the task is to persuade public administrations that it is a good strategy. Are you optimistic about the chances that public administrations will follow the recommendations of your Authority or do you see too much resistance?

A. The objective nature of corruption prevention measures hinges upon the capabilities for implementation on the part of public administration organizations. Bureaucratic—in the pejorative sense of the word—implementation does not help: it is useless. We need to be convinced of "the substance", of the spirit of corruption-prevention measures. Public administration organizations are not investing in corruption prevention. Our prevention measures are seen as a burden. They do not understand there is a new spirit underlying our measures: public administration organizations become an active subject for action, rather than a passive object of penal investigation.

Q. Is there a real chance results will come too much in the long run and in the meantime expectations of success are frustrated?

A. Absolutely. Yes, there is such a risk. Whatever happens now, people come out and say it is the Authority's fault. It must be made very clear that the Authority only does administrative prevention. On the other hand, this country is not accustomed to non-visible results. Whereas penal action is visible.

Q. Among the measures against corruption you tirelessly mention transparency. This is a major change which appears to not be shared by public administration and other actors (for instance, private entities under public control). Do you think there is actual and deliberate resistance? And how are you going to overcome that?

A. Transparency is the real antidote because transparency changes the notion of control. Through transparency, control becomes diffused, on the part of the citizens. If we do not start doing transparency, it will be very difficult to ask the citizens to do control. Transparency indeed is a burden on public administration we cannot do without. We do have resistance from public organizations: on the one hand, there is work to be done, and that is the origin of one type of resistance; on the other hand, there is a kind of embedded secrecy which is going to be very difficult to dismantle.

Q. What are you going to do about such resistance?

A. We challenge public administration organizations' attempts to be outside the perimeter of Authority supervision. We have been doing this even in court, with some success.

Q. Many—among the opposition to the new anti-corruption strategy— bewail the anti-corruption measures as burdening with red tape and rigid procedures public administrations that are already inefficient. How do you answer that criticism? How is it possible to reconcile the quest for impartial public administration and impartial managers with the quest for efficiency? Can anti-corruption plans be integrated with instruments of performance measurement?

A. The notion of the burden the Authority imposes on public organizations is unfounded. First of all, Authority regulation is not about the individual actions of public administration, therefore there is no reason Authority regulation should slow down the efficiency of public administration. The foregoing criticism is an end in itself: it says "no" only to say no. It shows people do not know what they are talking about. Second, impartiality on the part of public administration is efficient. Impartiality reduces litigious- ness. Third, there is certainly some learning to be done and the new functions that are asked of public organizations should be accepted as a challenge. It should be a sort of self-cleaning process.

Speaking of anti-corruption plans to be integrated with instruments of performance measurement—maybe the converse should be true: trans- parency should be included within the objectives of a public organiza- tion, rather than the objectives to be included in the elements of transparency. More than talking about bonuses based on performance, we should acquire the conviction that the Three-Year Plans to Prevent Corruption are a useful tool.

Q. What makes you think public administration organizations are interested in accepting this challenge?

A. We do not think they will accept the challenge because they are good; we are challenging them by showing they can do something different and we say this in public. We rely on the tried-and-tested name-and-shame mechanism.

On the role of the Authority

Q. The establishment of the Authority is perceived as an important innova- tion, in Italy and abroad. How much of this perception is due to an in- depth knowledge of the Authority's role and how much is only a superficial response?

A. Both elements are present. On the one hand, people do not have a clear idea of what the Authority is about. They think we are a kind of special public prosecutor. The Authority has displaced the traditional idea the only possibility is penal action. This is a new message: it is not only the judiciary that can work on corruption. Abroad there is a lot of attention on our work from international organizations like the OECD and GRECO. We are undertaking an autonomous process which is different from previous work done on corruption in the international arena. Some of the characteristics of the Authority appear to be unique to the Italian experience. These include:

- Independence of the Authority, both from the classical three Montesquieu powers: the executive, from parliament and of course the judiciary;
- Transparency on public procurement;
- Use of administrative police;
- Self-financing (through an ad hoc tax on public procurement).

On the other hand, there is scepticism about our plurality of functions. As I said, the Authority also works on transparency and public procurement, which is outside other authorities' remits abroad. Some anti-corruption organizations see the Authority as a possible future model. For instance, France's Service Central de Prévention de la Corruption 2015 Report has a whole chapter on the Authority. Also, Italy won two twinning bids within the EU: Serbia and Montenegro are paired with Italy in their path towards the Union.

Q. What is the diversity of the anti-corruption approach in the countries you have seen in your recent experience? Is there a short message you would like to send to your counterpart in some distant anti-corruption authority across the globe?

A. We need to create a network, to link together. There are too many differences among the anti-corruption efforts of different countries around the world. We need to develop a common denominator that is shared by a number of countries. On the other hand, we see a very kaleidoscopic landscape: some anti-corruption organizations are very sectoral; some have judicial power; not all of them do transparency; not all of them do prevention; not all of them do public procurement.

Q. In Italy there is a pendulum swing between great expectations about the Authority's work and devaluation of its powers and capability. Among the Authority's functions—regulation, supervision, recommendations, sanctions—which one you think is most important?

A. I can say for sure which one is the least important: that is the Authority's sanctioning power. The other three functions are key in a modern view of anti-corruption. The Authority does not do police work—inspections—we do orientation.

Q. What in your judgement is the key task of the Authority vis-à-vis public administrations?

A. The Authority supervision function is not a synonym for inspections. We are not a police force. The Authority has coined an oxymoron: "collaborative supervision". We do mentorship (*affiancamento*); we do coaching.

On the organization of the Authority

Q. The Authority is the result of a merger between previously existing organizations: CIVIT and AVCP. You proposed a restructuring plan and implemented a reorganization of authority bureaux. Do you think the merger was successful? Or maybe there is a disequilibrium in the distribution of tasks related to the key functions of the Authority: procurement, anti-corruption and transparency?

A. The merger did take place: a large organization (AVCP: Authority for the Supervision of Public Contracts) absorbed a small organization (CIVIT: Committee on Evaluation, Integrity and Transparency of Public Administration). It was a kind of reverse acquisition. Undoubtedly, there is—today, within the Authority—a strong skill base on procurement of public works. The numbers of this activity are larger, the skills of the personnel are oriented that way, and our tasks on public procurement are our most relevant tasks. This is a physical fact: procurement of public works is the biggest risk for corruption.

Q. What were the key difficulties of your reorganization effort?

A. We have had many difficulties. The former authority on public procurement of public works (AVCP, the biggest organization we come from) had a different core business from the present Authority. It was designed to do administrative supervision; it did not do anti-corruption. They did not have a comprehensive strategy. There was no integrated view of the corruption issue.

Q. What are the most obvious constraints of the organizational machine you are leading?

A. The previous organization we come from had an excessive burden of personnel which manifested itself—among other things—as a large number of managers. Being a manager was meant as a personnel qualification to raise people's salaries; it did not involve managerial tasks or capabilities.

Q. Do you think you can overcome these difficulties and these constraints? How are you going to do that?

A. We are trying, now that we can invest a little. We have some budget capability. We are getting young people onto our staff. We are implementing a flexible organizational structure. Previously, they had six directorate generals. Today we have a very horizontal and lean organizational structure.

Q. How would you like the activity of the Authority to be measured? Is there a specific indicator you would like to improve as a measure of Authority success in the containment of corruption?

A. As I said, we need to develop indices to measure our corruption containment effort. For instance, we could work on the number of complaint reports received vis-à-vis the number we actually worked on. Moreover: the number of investigations started and the number of investigations completed during a given period of time.

On the other hand, the OECD has appreciated our approach and our working method on the 2015 Milan Expo: that is certainly an indicator of success.[1] Also, the government and parliament call on us on a number of issues, and almost every week I participate in a hearing in parliament. That is also a sign of success. Corruption containment is being grounded in our social consciousness.

We need to broaden our view as well. It would be great to receive acknowledgement of our role by the citizens. It would be good to be accepted by the citizens in our role of prevention rather than repression.

Q. Do we go back to opinion polls?

A. We cannot work only through our formal and technical activities; we also need to work on our more general role. I would like to convince the people that you can have an authority with an important role without arresting anybody.

On the President's management style

Q. You are very popular. This comes from your reputation for integrity as a judge when you fought organized crime and corruption. However, this also comes from your communication ability. Is this a personal and irreplaceable skill of yours, or is there a way to make permanent the Authority's capability for communication and guide the addressees of your message?

A. I wouldn't know. Anyway, it is important to communicate when you do the kind of work that we are doing. I do not have any specific method. It is important to understand what the media want, what the media understand. We are experimenting. It has been the case that we did a lot of work and they did not care. We did a tremendous amount of work on the Three-Year Plans to Prevent Corruption—and the media did not care. The media listen to us when we do something specific. They like that. We are not setting priorities. They do. We need time to be able to define what is communicable material.

Q. You have a very good relationship with the press and the media. Is this fundamental in ensuring the success of your Authority?

A. This is indispensable. This is a must. The force of an authority is based on its distinctive character (*riconoscibilità*). We need the citizens. We need civic control.

Q. What is the risk of media over-exposure? What could be the effects: excessive expectations; a cult of personality within the Authority; professional jealousy; the erosion of the responsibility of managers in public administration and within the Authority itself? Will people think, "Cantone will take care of that"?

A. Excess media exposure is the highest risk. Our function is to make people in public administrations work on their own; our function is not to work on their behalf. We need to involve individual citizens. We are trying to clear a minefield. We can't do this in the short run. The President of the Authority was exposed because he had a previous franchise with the media. He had a previous identity (*riconoscibilità*) with them. Nonetheless, within the Authority there is full collegial management: the council of the Authority—its governing body—works in a unanimous way.

On external relations

Q. The Authority is an independent entity, tasked with regulation and supervision. However, it is also a key player in public administration and business, for instance in the area of public procurement. The Authority, therefore, has many partners: parliament, the cabinet, public administration, business and civil society. How can you establish a constructive relationship with each of them?

A. This is the hardest task. It takes a lot of energy. I meet a wide variety of groups. We try to talk to everybody. All channels are open, for instance with central government, local authorities, universities. We are trying to be open.

Q. Let us review what worked and what went wrong in your relationship with your key publics: parliament, the cabinet, the polity (the political system), public administrations, business and civil society.

A. Our relationship with parliament did work OK. They have always been open with us and have listened to us. They did not always do what we said. However, they always heard us.

The cabinet has kept us in the loop. Our medal is our involvement with the 2015 Milan Expo.

We have had relationships with all political forces. We are trying to be impartial vis-à-vis political forces and they seem to appreciate that.

Our relationship with public administrations is the most problematic area. Many public administration organizations pretend they are listening to us. Some of them are authentic in their relationship with us. They also understand we are an element of public administration ourselves.

We have a good dialogue with business. They do not always accept our intervention, especially when their relationships with public administration are long established. Interestingly, however, the strongest request for Authority powers—in the reform of the Public Procurement

Code—came from private economic parties: they were happy to have one place they could talk to.

We have a decent relationship with civil society. We participate in many public initiatives. We have signed protocols with a number of associations, like Libera [a Sicilian initiative] and Transparency International. We work with the Ministry of Education. We are concerned they might expect too much from us and we do not want to let them down.

Conclusions

Q. You and your authority are halfway through your six-year mandate. What would you like to be your legacy to parliament and the cabinet by the end of your mandate, in 2020?

A. Now I would say: Let us go forward. Let us go beyond emergency logic. Let us keep going when the spotlight is turned off.

Q. What is the Italy you found when you came in and how would you like to leave it when your mandate is over?

A. I would like to leave better awareness of the problem of corruption among the citizens. There is not enough of that.

Q. What difference would you like to have made in the Italian administration system (in the wider sense)?

A. I would like to leave closer attention to transparency and rules. More sensitivity to rules even in situations that are not at risk of corruption. We are very careful that our rules are not going to be burdensome and are effective in maintaining the impartial action of public administration. We do a lot of prior consultation to ensure the rules are effective. Effectiveness of rules is, in fact, not guaranteed by their legitimacy.

Q. In hindsight, imagine you were starting again tomorrow morning: what would you not do, this time?

A. I wouldn't know. Three years is too little time to tell. Maybe we should have made the whole thing less personal. This could become a boomerang after the Renzi Cabinet left office [at the end of 2016]. The Renzi Cabinet appointed us and we have the task of making our work stable, not so much dependent on a specific cabinet. There is a risk from the political side; however, the Renzi Cabinet never interfered with our activities. We have a new cabinet now [the Gentiloni Cabinet] and we have not experienced any changes in our relationship with the premier and his ministers.

Q. The media are saying that now the Renzi Cabinet has left office the Authority runs the risk of becoming just another ministry.

A. Well, this is our task: to try to make this not the case, through our record of impartial action. Recognition that we are perceived as impartial is key for our policies to continue on a bipartisan basis.

We would like to summarize Raffaele Cantone's point of view as one in which he emphasizes the massive scale of corruption in Italy. We agree that perception is very high, but nonetheless Cantone is aware of the pervasive character of corruption in Italy. It may be worth repeating that he has a very broad definition of corruption, to include not only "technical" corruption, in the penal sense, but "maladministration" in general.

Besides public administration organizations, Cantone placed a high emphasis on the relationship with the citizens. Anti-corruption action should engage the citizens and not be limited to an "intra-administration" business.

Note

1 The Authority played a key role in preventing further cases of corruption and at the same time important international deadlines were met. See the Expo case history in the text.

Glossary

Access, civic accesso civico
Accountability responsabilità as in "corruption equals monopoly plus discretion minus accountability"/rendicontazione, responsabilità dovuta, responsabilità ottemperata
Act legge
Activities, cyclical attività ricorrenti
Advice, legal parere
Appraisal of modification (in procurement evaluation) perizia di variante
Assistance in legal investigation soccorso istruttorio
At request of the office/Of its own initiative iniziativa d'ufficio
Audit committee collegio dei revisori dei conti
Auditors, Court of Corte dei Conti
Authority, for Public Contracts, Supervision Autorità vigilianza sui contratti pubblici (AVCP)
Award of contracts affidamento
Benchmark price prezzi di riferimento
Bid evaluation committees, register of members of membri delle commissioni di gara
Bid rigging turbativa d'asta
Bribe price list prezzario delle tangenti
Building permits permesso a costruire
Call for tender bando di gara
Capabilities, present attuali capacità e potenzialità
Case Specific Supervision Vigilanza Puntuale
Central Purchasing Body Organismo di aggregazione
Certification Bodies Società Organismo di Attestazione (SOA)
Certification System, Supervision of the attività di vigilanza sul sistema di qualificazione
Civic Access vedi Access
Closing memo nota di chiusura
Commissioner Commissario

Company rating   rating di impresa—about each company's past technical performance

Company under public control   società in controllo pubblico

Comparative Public Administration   Pubblica Amministrazione Comparata

Competitive bid   gara pubblica

Competitive bidding   gara d'appalto

Compliance   verifica adempimenti

Composition with the creditors   concordato preventivo

Contracting authorities   Stazioni Appaltanti (SA)

Contracting awarding body   stazione appaltante

Contractor, general   contraente generale

Corporate Social Responsibility   Responsabilità Sociale delle Imprese

Corporations and other bodies that are only participated (minority share) by public administration organizations   società a partecipazione pubblica

Corruption   corruzione

Corruption Prevention Officers   responsabile della prevenzione della corruzione e della trasparenza

Counsel   parere [vedi anche sopra]

Decision   delibera

Decision to launch a procurement process   determina a contrarre

Decree, Law   Decreto Legge (d.l.)—an act written by the executive branch, needing to be made a full act by the parliamentary assembly within 60 days

Decree, Legislative   Decreto Legislativo (d.lgs.)—an act written by the executive branch and passed only by parliamentary commissions, not by the parliament assembly

Department of Public Service   Dipartimento della Funzione Pubblica

Determination of contracting   determina a contrarre

Direct award contract   affidamento diretto

Discharge   rescissione

Discharge by agreement   rescissione consensuale

Discretion   discrezione, discrezionalità nella gestione

Draw   sorteggiato

Economic unit (company)   operatore economico

End terminals   terminali

Ex parte motion   iniziativa ad istanza di parte

Exercise of sanctioning power   esercizio del potere sanzionatorio

Extension, of contracts   proroga

External appointment   Incarichi Esterni

Extraordinary administration   Aministrazione Straordinaria

Fine   sanzione con multa

Fine on contract   sanzioni sul contratto

Fragmentation of procurement orders frammentazione dell'appalto

Functionaries (middle-level employees) funzionari

Functions of regulation Funzioni di Regolazione

Good bene

Guideline Linea Guida

Guidelines of an Inter-Ministries Committee Linee di indirizzo del Comitato Interministeriale

Hospital corporation azienda ospedaliera

Implementation a detailed description of actions taken, leading to an understanding of the possible planned and unintended consequences of such actions

Independent Commission for the Evaluation, Integrity and Transparency Commissione Indipendente per la Valutazione, l'Integrità e la Trasparenza (CIVIT)

Independent Evaluation Unit Organismi Indipendenti di Valutazione (OIV)

Inspections attività ispettive

Integrated Tender Appalto Integrato

Interim relief soccorso istruttorio

Interpretation Acts atti interpretativi

Investigation accertamento

Judgement sentenza

Jurisdictional litigation contenzioso giurisdizionale

Law diritto

Law Decree see *Decree, Law*

Legislative Decree see *Decree, Legislative*

Local Health Units Aziende Sanitarie Locali (ASL)—public organization that organizes health and hospital services

Maladministration cattiva amministrazione or malamministrazione

Manager (High-Level Employees) dirigente

Managerial sensitivity sensibilità gestionale

Ministry of Industry and Economic Development (MISE) Ministero dell'Industria e dello Sviluppo Economico

Ministry of the Economy and Finance (MEF) Ministero dell'Economia e delle Finanze

Misconduct condotta illecita

Mission Oriented Unit Unità o Struttura di Missione

Model of calls for tender Bando Tipo

Monetized quantity of yearly corruption in Italy entità monetizzata della corruzione annuale in Italia

Monopoly monopolio—corruption = monopoly + discretion − accountability

Motion istanza

Municipality comuni

National Anti-corruption Plan Piano Nazionale Anticorruzione (PNA)

National Health Care Service Servizio Sanitario Nazionale (SSN)

Observatory of public contracts for works, services and supplies osservatorio dei contratti pubblici relativi a lavori, servizi e forniture

Organizational behaviour comportamento organizzativo

Organizations governed by private law organismi di diritto privato

Organizations only partially owned by public administration/simple public shareholding società in controllo o a partecipazione pubblica

Parliamentary Questions Interrogazioni Parlamentari

Pre-litigation pre-contenzioso

Prevention prevenzione

Procurement contract contratto pubblico

Procurement of goods and services acquisto di beni e servizi

Procurement of services appalto di servizi

Procurement unit unità per acquisti

Procurement, public appalto pubblico

Project modification (or changes) varianti progettuali

Public company società in mano pubblica

Public corporation impresa o ente pubblico

Public Law Legge

Public Procurement and Concession Contract Code—Legislative Decree no. 50 of 2016 Codice degli Appalti—Decreto Legislativo no. 50 del 2016

Public work lavoro pubblico

Relevant market mercato di riferimento

Renegotiation of contract revisione del contratto

Renewal of contract proroga contrattuale

Revolving doors vedi *pantouflage*

Sanctions affairs sanzioni

Service servizio

Social Responsibility Responsabilità Sociale

Special Supervision attività di vigilanza straordinaria/controllo

Specification of contract capitolato d'appalto

Standard cost costi standard

Supply fornitura

Suspension of contract sospensione del contratto

Tax Law Decreto Fiscale

Tender Notice Bando di Gara

Tender, public gara pubblica, offerta formale, gara d'appalto

Tendering procedure gara d'appalto

Tendering Unit Stazione Applatante

Termination of Contract Risoluzione Contrattuale

Terms of Contract   Capitolato
Territorial scope   ambito territoriale
Three-Year Plan   Piano Triennale
Three-Year Plan to Prevent Corruption and Transparency   Piano Triennale per la Prevenzione della Corruzione e la trasparenza (PTPCT)
Three-Year Plan to Prevent Corruption   Piano Triennale per la Prevenzione della Corruzione (PTPC)
Transparency   trasparenza
Work(s)   lavori

Bibliography

Ackerman, B. (2000). The new separation of powers. *Harvard Law Review*, 113, 633–729.

Andracchio, D. (2016). Il divieto di "pantouflage": Una misura di prevenzione della corruzione nella pubblica amministrazione. *Rivista di Diritto Amministratovo*, 9. Retrieved from: www.GiustAmm.it.

Anthony, R.N., and Herzlinger, R.E. (1975). *Management Control in Nonprofit Organizations*. Chicago: R.D. Irvin.

Arena, G. (2008). Le diverse finalità della trasparenza amministrativa. In F. Merloni (Ed.), *La Trasparenza Amministrativa*, pp. 29–43. Milan: Giuffré.

Australian Collaboration. (2013). *Democracy in Australia: Corruption*. Retrieved 21 December 2017 from: www.australiancollaboration.com.au/pdf/Democracy/Anti-corruption-commissions.pdf.

Banfield, E.C. (1967). *The Moral Basis of a Backward Society*. New York: Free Press.

Bertot, J.C., Jaeger, P.T., and Grimes, J.M. (2010). Using ICTs to create a culture of transparency: E-government and social media as openness and anti-corruption tools for societies. *Government Information Quarterly*, 27, 264–71.

Birchall, C. (2014, February). Radical transparency? *Critical Methodologies*, 14(1), 77–88.

Birkinshaw, P. (2006). Freedom of information and openness: Fundamental human rights? *Administrative Law Review*, 58(1), 177.

Brandeis, W. (1914). *Other People's Money: And How the Bankers Use it*. Retrieved 1 August 2018 from: www.law.louisville.edu.

Broussolle, Y. (2017). Les principales dispositions de la loi Sapin pour la transparence et la modernisation de la vie économique. *Gestion & Finances Publiques*, 2, 108–13.

Cantone, R. (2016). The new Italian Anticorruption Authority: Duties and perspective. *Digest: National Italian American Bar Association Law Journal*, 83–100.

Cantone, R., and Merloni, F. (2015). *La Nuova Autorità Nazionale Anticorruzione*. Turin: G. Giappichelli Editore.

Carloni E. (2014). *L'amministrazione aperta. Regole, strumenti, limiti dell'Open Government*. Rimini: Maggioli.

Carloni, E. (2017). Misurare la corruzione? Indicatori di corruzione e politiche di prevenzione. *Politica del Diritto*, 3.

Carloni, E. (2018). Italian anticorruption and transparency policies: Trends and tools in combating administrative corruption. In A. Grasse, M. Grimm, and J. Labitzke (Eds.), *Italien zwischen Krise und Aufbruch*, pp. 365–86. Berlin: Springer.

Carloni, E., and Giglioni, F. (2017). Three transparencies and the persistence of opacity in the Italian government system. *European Public Law*, 23(2), 285–99.

Cerrillo Martinez, A. (2011). The regulation of diffusion of public sector information via electronic means: Lessons from the Spanish regulation. *Government Information Quarterly*, 28(2), 188–202.

Chandler, A.D. (1990). *Strategy and Structure: Chapters in the History of the Industrial Enterprise* (Vol. 120). Cambridge, MA: MIT Press.

Cordis, A.S., and Warren P.L. (2014). Sunshine as disinfectant: The effect of state Freedom of Information Act laws on public corruption. *Journal of Public Economics*, 115, 18–36.

Cottarelli C. (2014, 27 March). *Proposte per una revisione della spesa pubblica 2014–2016: Commissario straordinario per la revisione della spesa*. Rome: Presidenza del Consiglio dei Ministri, Roma.

Cottarelli, C. (2016). *La lista della spesa: La verità sulla spesa pubblica italiana e su come si può tagliare*. Milan: Feltrinelli Editore.

Cuillier, D., and Piotrowski, S.J. (2009). Internet information-seeking and its relation to support for access to government records. *Government Information Quarterly*, 26(3), 441–9.

D'Anselmi, P., Chymis, A., and Di Bitetto, M. (2017). *Unknown Values and Stakeholders: The Pro-Business Outcome and the Role of Competition*. London: Palgrave.

David, D. (2017). Lobbying, gruppi di interesse e regolazione amministrativa. *Rivista di Diritto Pubblico Italiano, Comparato, Europeo*, 24. Retrieved from: www.federalismi.it.

Della Porta, D., and Vannucci, A. (2005). The moral (and immoral) costs of corruption. In U. Von Alemann (Ed.), *Dimensionen politischer Korruption*, pp. 109–34. Wiesbaden: Springer Verlag.

Di Mascio, F., and Natalini, A. (2013). Context and mechanisms in administrative reform processes: Performance management within Italian local government. *International Public Management Journal*, 16(1), 141–66.

Donato, L. (Ed.) (2016, February). *La riforma delle stazioni appaltanti: Ricerca della qualità e disciplina europea*. Quaderni di Ricerca Giuridica della Consulenza Legale, 80.

Eccles, R.G., and Krzus, M.P. (2010). *One Report: Integrated Reporting for a Sustainable Strategy*. New York: John Wiley.

Expo Milano 2015. (2015). *The Protocol of Understanding between Anti-Corruption Expo Milano 2015, ANAC and OECD has been Presented*. Retrieved on 14 July 2018 from: www.expo2015.org/archive/en/news/all-news/the-protocol-of-understanding-between-anti-corruption-expo-milano-2015--anac-and-oecd-has-been-presented.html.

Fields, W.S. (1994). The enigma of bureaucratic accountability. *Catholic University Law Review*, 43, 505–6.

Fiorino, N., and Galli, E. (2013). La corruzione in Italia. *Il Mulino*. Retrieved from: www.mulino.it.

Fiorino, N., Galli, E., and Petrarca, I. (2012). Corruption and growth: Evidence from the Italian regions. *European Journal of Government and Economics*, 1(2), 126–44.

Fiorino, N., Galli, E., and Petrarca, I. (2013). Press and corruption in the Italian regions: An empirical test. Retrieved from: www.siecon.org/online/wp-content/uploads/2014/10/Fiorino-Galli-Petrarca-83.pdf.

France Soir (2017, 7 November). L'Agence française anticorruption mise sur la prévention. Retrieved from: www.francesoir.fr/actualites-economie-finances/lagence-
francaise-anticorruption-mise-sur-la-prevention.

Franzini, M., Granaglia, E., and Raitano, M. (2016). *Extreme Inequalities in Contemporary Capitalism: Should We Be Concerned About the Rich?* New York: Springer.

Galetta, D.U. (2014). Transparency and access to public sector information in Italy: A
proper revolution? *Italian Journal of Public Law*, 2, 213–31.

Galli, E. (2013, 28 November). *Analisi Economica della Corruzione nella Pubblica Amministrazione: Cause ed Effetti* ("Economic analysis of corruption in public administration: causes and effects"). Università degli Studi, L'Aquila Ministero dell'Economia
e delle Finanze Scuola Superiore dell'Economia e delle Finanze Roma.

Gardini, G. (2014). Il codice della trasparenza: Un primo passo verso il diritto
all'informazione amministrativa? *Giornale di Diritto Amministrativo*, 8–9, 875–91.

Giarda, P. (2012, 24 January). Dinamica, struttura e criteri di governo della spesa pubblica: Un rapporto preliminare. *XVI Legislatura*. Interventi spesa pubblica. Retrieved
from: http://leg16.camera.it.

Golden, M., and Picci, L. (2006). Corruption and the management of public works
in Italy. In S. Rose-Ackerman (Ed.), *International Handbook on the Economics of Corruption*, pp. 457–83. Cheltenham, UK: Edward Elgar.

Gray, C.W., and Kaufman, D. (1998). Corruption and development. *Finance and
Development*, 35(1), 7–10.

Heady, F. (2001). *Public Administration: A Comparative Perspective* (6th ed.). Abingdon,
UK: Routledge (1st ed.: Prentice Hall, 1966).

Heald, D. (2006). Transparency as an instrumental value. In C. Hood and D. Heald
(Eds.), *Transparency: The Key to Better Governance?* Oxford: Oxford University Press.

Il Giornale (2016, 12 October). Retrieved from: www.ilgiornale.it/.

Johnston, M. (2012). Corruption control in the United States: Law, values, and the
political foundations of reform. *International Review of Administrative Sciences*, 78(2),
329–45.

Kierkegaard, S. (2009). Open access to public documents: More secrecy, less transparency! *Computer Law & Security Review*, 25(1), 3–27, https://doi.org/10.1016/j.
clsr.2008.12.001.

Klitgaard, R. (1988). *Controlling Corruption*. Los Angeles: University of California
Press.

Klitgaard, R. (1997, November). *International Cooperation Against Corruption*. Manuscript.

Klitgaard, R., (2014, December). *Controlling Corruption Together*. Manuscript, Clare-
mont Graduate University.

Kreimer, Seth F. (2008, June). The Freedom of Information Act and the ecology of
transparency. *University of Pennsylvania Journal of Constitutional Law*, 10(5), 1011–80.

La Noce, M. (2015). Designing a management information system for competition
law agencies. In M. Di Bitetto, A. Chymis and P. D'Anselmi (Eds.), *Public Management as Corporate Social Responsibility: The Economic Bottom Line of Government*,
pp. 11–26. New York: Springer.

Lapiccirella, A. (2015). On bureaucratic behavior. In M. Di Bitetto, A. Chymis, and
P. D'Anselmi (Eds.), *Public Management as Corporate Social Responsibility: The Economic Bottom Line of Government*, pp. 103–18. New York: Springer.

Light, P.C. (1993). *Monitoring Government: Inspectors General and the Search for Accountability*. Washington, DC: Brookings Institution Press.

Merloni, F. (2006). Dirigenza pubblica e amministrazione imparziale: Il modello italiano in Europa. *Il Mulino*.

Merloni, F. (Ed.) (2008). *La trasparenza amministrativa*. Milan: Giuffré.

Merton, R.K., Gray, A.P., Hockey, B., and Selvin, H.C. (1952). *Reader in Bureaucracy*. New York: The Free Press.

Miller, T., Kim, A.B., and Robert J.M. (2018). *Index of Economic Freedom*. Washington, DC: Heritage Foundation.

Mitchell, R.K., Agle, B.R., and Wood, D.J. (1997). Toward a theory of stakeholder identification and salience: Defining the principle of who or what really counts. *Academy of Management Review*, 22(4), 853–86.

Newell, J.L., and Bull, M.J. (2003). *Corruption in Contemporary Politics*. London: Palgrave Macmillan.

Olson, M. (1965). *Logic of Collective Action: Public Goods and the Theory of Groups*. Harvard Economic Studies, Vol. 124. Cambridge, MA: Harvard University Press.

Olson, M. (2008). *The Rise and Decline of Nations: Economic Growth, Stagflation, and Social Rigidities*. New Haven, CT: Yale University Press.

Panorama (2012). Il costo della corruzione in Italia: 60 miliardi di euro all'anno. *Panorama*. Retrieved 8 April 2015 from: http://archivio.panorama.it/economia/Il-costo-della-corruzione-in-Italia-60-miliardi-di-euro-all-anno.

Parisi, N. (2017, 22 December). Verso l'emersione di un modello internazionale di prevenzione della corruzione. *Rivista di Diritto Pubblico Italiano, Comparato, Europe*. Retrieved from: www.federalismi.it/nv14/articolo-documento.cfm?Artid=35368& content=Verso+l%27emersione+di+un+modello+internazionale+di+prevenzione +della+corruzione&content_author=%3Cb%3ENicoletta+Parisi%3C/b%3E.

Pascale, R.T. and Athos, A.G. (1981). *The Art of Japanese Management*. New York: Simon & Schuster.

Posner, E.A., and Vermeule A. (2010). *The Executive Unbound: After the Madisonian Republic* (1st ed.). Oxford, UK: Oxford University Press.

Putnam, R. (1993). *Making Democracy Work: Civic Traditions in Modern Italy*. Princeton, NJ: Princeton University Press.

Putnam, R. (2016, 26 September). *Our Kids: The American Dream in Crisis. Simon & Schuster Regions: An Empirical Test*. Presented at SIEP Conference, Pavia, Italy.

Quah, J.S.T. (2009). Benchmarking for excellence: A comparative analysis of seven Asian anti-corruption agencies. *Asia Pacific Journal of Public Administration*, 31(2), 171–95.

Repubblica (2016, 19 November). Torino, Parisi all'attacco di Cantone: "E' una iattura per il Paese." Retrieved from: http://torino.repubblica.it/cronaca/2016/11/19/news/ torino_parisi_all_attcco_di_cantone_e_una_iattura_per_il_paese_-152335204.

Richards, D., and Smith, M.J. (2015). In defence of British politics against the British political tradition. *The Political Quarterly*, 86(1), 41–51.

Ricolfi, L. (2007). *Le tre società. É ancora possibile salvare l'unità d'Italia?* Milan: Guerini e Associati.

Roberts, A. (2006). Governmental adaptation of transparency rules. In C. Hood and D. Heald (Eds.), *Transparency: The Key to Better Governance?* Oxford, UK: Oxford University Press.

Rose-Ackerman, S. (1996). Altruism, nonprofits, and economic theory. *Journal of Economic Literature*, 34, 701–28.

Rose-Ackerman, S., and Palifka, B.J. (2016). *Corruption and Government: Causes, Consequences, and Reform*. Cambridge, UK: Cambridge University Press.

Savino, M. (2013). Le norme in materia di trasparenza amministrativa e la loro codificazione. In B.G. Mattarella and M. Pelissero (Eds.), *La legge anticorruzione*, pp. 113–23. Turin: Giappichelli.

Schelling, T. (1978). *Micromotives and Macrobehavior*. New York: W.W. Norton.

The Economist (2014a, 9 August). High-class errand boys: Parliamentary workers are facing a cut in their generous pay.

The Economist (2014b, 9 August). Mandarin lessons: Governments need to rethink how they reward and motivate civil servants.

The Economist (2015, 14 November). Aiwa (yes) minister: The region's countries desperately need to reform their public sectors.

The Economist (2016a, 2 January). From red tape to joined-up government: Latin America's efforts to improve public policies are often undermined by politicised and obsolete civil services.

The Economist (2016b, 1 April). Corruption: Bad press does Europe a disservice.

The Economist (2016c, 26 November). Why Italy should vote no in its referendum: The country needs far-reaching reforms, just not the ones on offer.

Thompson, D. (1992). Paradoxes of administrative ethics. *Public Administration Review*, 52(3), 254–9.

Tuzi, Fabrizio (2016). *Amministrazione Pubblica: Dall'egoismo alla competizione*. Soveria Mannelli: Rubbettino.

Vannucci, A. (2013). *La corruzione in Italia: Cause, dimensioni, effetti*. In B.G. Mattarella and M. Pelissero (Eds.), *La legge anticorruzione*, pp. 25–58. Turin: Giappichelli.

Wilson, W. (1887, July). The study of administration. *Political Science Quarterly*.

Comparative international and supranational sources

European Central Bank (2016, August). *The fiscal and macroeconomic effects of government wages and employment reform* (J.J. Pérez et al.). European Central Bank, Eurosystem Occasional Paper Series, 176.

European Commission (2012, February). Eurobarometer 76.1, TNS, Opinion and Social (2012) 374. *Corruption Report*. Retrieved from: http://ec.europa.eu/commfrontoffice/publicopinion/archives/ebs/ebs_374_en.pdf.

European Commission (2014). *Corruption Report*. Requested by the Directorate-General for Home Affairs and co-ordinated by the Directorate-General for Communication.

European Commission (2014, February). Special Eurobarometer 397/Wave EB79.1, TNS Opinion and Social Fieldwork: February–March 2013. Retrieved from: http://ec.europa.eu/public_opinion/index_en.htm.

European Commission (2014, 3 February). *Report from the Commission to the Council and the European Parliament. EU Anti-Corruption Report*. Brussels. COM(2014) 38 final.

European Council, GRECO (Group of Countries against Corruption). Joint first and second evaluation round, evaluation report on Italy, 2 July 2009. 1st evaluation round: Independence, specialization and means available to national bodies engaged in the prevention and fight against corruption; Extent and scope of immunities. 2nd evaluation round: Proceeds of corruption; Public administration and corruption; Legal persons and corruption.

European Council, GRECO (Group of Countries against Corruption). Compliance report on Italy, 23–27 May 2011.

European Council, GRECO (Group of Countries against Corruption). Third evaluation round, evaluation report on Italy. Theme I: Incriminations; Theme II: Transparency of party funding; 20–23 March 2012.

European Council, GRECO (Group of Countries against Corruption). Addendum to the Compliance report on Italy, 17–21 June 2013.

European Council, GRECO (Group of Countries against Corruption). Compliance report on Italy, 16–20 June 2014.

European Council, GRECO (Group of Countries against Corruption). Fourth evaluation round. Themes examined: prevention of corruption in respect of members of parliament, judges and prosecutors.

European Parliament 2014–2019 (2016). *The Cost of Non-Europe in the Area of Organised Crime and Corruption. Annex II: Corruption,* EPRS (European Parliamentary Research Service) European Added Value Unit PE 579.319. Paper by RAND Europe. March 2016.

European Parliament 2014–2019 (2016). *The Cost of Non-Europe in the area of Organised Crime and Corruption, Annex III: Overall Assessment of Organized Crime and Corruption,* EPRS (European Parliamentary Research Service) European Added Value Unit, PE 579.320. Briefing paper by Prof. Federico Varese. March 2016.

European Parliament 2014–2019 (2016, April). *Draft Report on the fight against corruption and follow-up of the CRIM resolution* (2015/2110(INI)). Committee on Civil Liberties, Justice and Home Affairs. Rapporteur: Laura Ferrara.

ICAC (Independent Commission Against Corruption) New South Wales (2017). *Annual Report 2016–2017.* Sydney, Australia.

ICAC (Independent Commission Against Corruption) New South Wales (2017, December). *Strategic Plan 2017–2021.* Retrieved from: www.icac.nsw.gov.au/documents/about-the-icac/4395-icac-strategic-plan-2017-2021/file.

IMF (International Monetary Fund) (2010, September). *Evaluating Government Employment and Compensation,* Benedict Clements, Sanjeev Gupta, Izabela Karpowicz, and Shamsuddin Tareq, Fiscal Affairs Department.

IMF (International Monetary Fund) (2016, May). *Fiscal Affairs and Legal Departments, Corruption: Costs and Mitigating Strategies.* Prepared by a staff team from the Fiscal Affairs Department and the Legal Department. SDN/16/05.

OECD (Organisation for Economic Co-operation and Development) (1999). *Public Sector Corruption: An International Survey of Prevention Measures.* OECD Publishing.

OECD (Organisation for Economic Co-operation and Development) (2008–2013). *Specialised Anti-Corruption Institutions. Review of Models.* Anti-Corruption Network.

OECD (Organisation for Economic Co-operation and Development) (2010). *The Right to Open Public Administrations in Europe: Emerging Legal Standards* (Ed. M. Savino). Sigma Paper, 46.

OECD (Organisation for Economic Co-operation and Development) (2013, 20 September). OECD Public Governance Reviews. *OECD Integrity Review of Italy: Reinforcing Public Sector Integrity, Restoring Trust for Sustainable Growth,* preliminary version. Retrieved February 2015 from: www.funzionepubblica.it.

OECD (Organisation for Economic Co-operation and Development) (2014). MSTI (Main Science and Technology Indicators), "Main Science and Technology Indicators 2015–1."

OECD (Organisation for Economic Co-operation and Development) WGB (Working Group on Bribery) (2014). *Annual Report*.

Quality of Government Institute (2010). *Measuring the Quality of Government and Sub-national Variations: A Dataset*. Retrieved from: www.qog.pol.gu.se/data/euproject.

Ragioneria generale dello Stato, Ministero dell'Economia e delle Finanze, Italy (2013). *Conto annuale del personale*.

Republic of Austria, Federal Ministry of the Interior, Federal Bureau of Anti-Corruption (BAK). www.bak.gv.at.

Service Central de Prévention de la Corruption, France (2015). *Rapport pour l'année 2015*. La documentation Française. Paris.

Statista (2016). Retrieved 28 November 2016 from: www.statista.com/statistics/269684/national-debt-in-eu-countries-in-relation-to-gross-domestic-product-gdp/.

Transparency International (2016). *UK Press Briefing: Unexplained Wealth Orders*. Retrieved from: www.transparency.org.uk/wp-content/uploads/2016/10/TI-UK-UWO-Press-Briefing-13-10-2016-1-1.pdf.

Transparency International (2018). *Corruption Perception Index 2018*. Retrieved from: www.transparency.org/cpi2018.

UNCAC (United Nations Convention Against Corruption) (2004). General Assembly Resolution 58/4 of 31 October 2003. United Nations Office on Drugs and Crime, Vienna/New York.

UNGC (United Nations Global Compact) Principle Ten: Anti-Corruption. Retrieved on 26 December 2017 from: www.unglobalcompact.org/what-is-gc/mission/principles/principle-10.

United Nations IRG (Implementation Review Group) (2016) of the United Nations Convention Against Corruption (UNCAC) of 2004. General Assembly Resolution 58/4 of 31 October 2003. UNODC (United Nations Office on Drugs and Crime), Vienna. Retrieved from: www.unodc.org/unodc/en/treaties/CAC/IRG-sessions.html.

United States Office of Government Ethics, *Agency Profile: Preventing Conflicts of Interest*. www.oge.gov.

United States Office of Government Ethics, *Standards of Ethical Conduct for Employees of the Executive Branch*. Retrieved 1 January 2017 from: www.oge.gov.

United States Office of Government Ethics (2015). *Conflict of Interest Prosecution Survey*. Retrieved 27 July 2016 from: www.oge.gov.

World Bank (2014a) *Doing Business 2015 Going Beyond Efficiency: Comparing Business Regulations for Domestic Firms in 189 Economies. A World Bank Group Flagship Report*.

World Bank (2014b). *WGI World Government Indicators*.

World Bank (2015a). WDI indicators. Retrieved 28 March 2018 from: http://databank.worldbank.org/data/reports.aspx?source=World-Development-Indicators.

World Bank (2015b). *Trust, Voice, and Incentives: Learning from Local Success Stories in Service Delivery in the Middle East and North Africa* (H. Brixi, E. Lust, and M. Woolcock). Washington, DC: World Bank.

World Bank (2017, 26 September). *Combating Corruption*. Retrieved 20 October 2017 from: www.worldbank.org/en/topic/governance/brief/anti-corruption.

World Economic Forum (2014). *The Global Competitiveness Report 2014–2015*. Retrieved from: http://www3.weforum.org/docs/WEF_GlobalCompetitiveness-Report_2014-15.pdf.

ANAC (Autorità Nazionale Anticorruzione) and other Italian government sources

(2009). *Relazione al Parlamento del SAeT (Servizio Anticorrupzione e Trasparenza)* ("Report to Parliament by the Anti-corruption and Transparency Service").

(2010). Anticorruzione_rapporto-SAET_2010/Dipartimento della Funzione Pubblica Autorità Nazionale Anticorruzione RELAZIONE Anno 2010—Servizio Anticorruzione e Trasparenza—SAeT.

(2011). *Relazione al Parlamento del SAeT.*

(2012). Rapporto della Commissione per lo studio e l'elaborazione di proposte in tema di trasparenza e prevenzione della corruzione nella pubblica amministrazione La corruzione in Italia. Per una politica di prevenzione, 2012.

(2013). *Corruzione sommersa e corruzione emersa in Italia: Modalità di misurazione e prime evidenze empiriche* ("Submerged Corruption and Evident Corruption in Italy: Measures and Initial Empirical Evidence"). ANAC.

(2013, 16 April). *Codice di comportamento dei dipendenti pubblici* ("Code of Conduct of Public Employees"). Decreto del Presidente della Repubblica no. 62.

(2013, 24 July). *National Anti-corruption Plan 2013 (PNA 2013 for years 2014–2016),* Piano Nazionale Anticorruzione—PNA, Legge 6 novembre 2012 n. 190 *Disposizioni per la prevenzione e la repressione della corruzione e dell'illegalità nella pubblica amministrazione,* Presidenza del consiglio dei ministri, Dipartimento della funzione pubblica, Servizio Studi e Consulenza Trattamento del Personale. Retrieved from: www.funzionepubblica.it.

(2013, December). Rapporto sul primo anno di attuazione della legge n. 190/2012.

(2014, December). Agreement with the Ministry of the Economy (Accordo con il MEF) on the range/perimeter of application of ANAC activity on controlled companies and participated companies (State Owned Enterprise SOE).

(2014, 30 December). "New Organization Plan" (Piano di riordino).

(2015, 4 July). Annual Report to Parliament for 2014.

(2015, 14 July). *1st meeting of the ANAC RPCs, managers for the prevention of corruption.*

(2015, 28 October). *Update of the PNA 2014–2016.*

(2016). *La spesa per redditi da lavoro dipendente: confronto tra Germania, Francia, Italia, Regno Unito e Spagna,* Ministero Economia e Finanze—Dipartimento della Ragioneria Generale dello Stato. Retrieved November 2016 from: www.contoannuale. tesoro.it.

(2016, 24 May). *2nd meeting of the Italian Authority with the Corruption Prevention Officers.*

(2016, 14 July). *Relazione Annuale al Parlamento dell'Autorità Nazionale Anticorruzione per l'anno 2015* ("Annual Report to Parliament for 2015").

(2016, 3 August). New National Anti-corruption Plan for the years 2016–2018. Retrieved from: www.anticorruzione.it.

(2017). *Resolution on Whistleblowing* ("Il whistleblowing in Italia, a cura di Anna Corrado", 22 June 2017).

(2017, 6 July). Annual Report to Parliament for 2016.

(2017, 22 November). *2017 Update to the National Anti-Corruption Plan (PNA) 2016–2018.*

Index

Page numbers in **bold** denote tables, those in *italics* denote figures.